Fuzzy Surfaces in GIS and Geographical Analysis

Theory, Analytical Methods, Algorithms, and Applications

Fuzzy Surfaces in GIS and Geographical Analysis

Theory, Analytical Methods, Algorithms, and Applications

Edited by
Weldon Lodwick

CRC Press
Taylor & Francis Group
Boca Raton London New York

CRC Press is an imprint of the
Taylor & Francis Group, an **informa** business

CRC Press
Taylor & Francis Group
6000 Broken Sound Parkway NW, Suite 300
Boca Raton, FL 33487-2742

First issued in paperback 2019

ISBN-13: 978-0-8493-6395-5 (hbk)
ISBN-13: 978-0-367-38799-0 (pbk)

Library of Congress Cataloging-in-Publication Data

Lodwick, Weldon A.
  Fuzzy surfaces in GIS and geographical analysis : theory, analytical methods, algorithms, and applications / Weldon Lodwick.
      p. cm.
  Includes bibliographical references and index.
  ISBN 978-0-8493-6395-5 (alk. paper)
  1. Geographic information systems--Mathematics. 2. Fuzzy mathematics. 3. Surfaces. I. Title.

G70.23.L63 2008
526.0285--dc22   2007032995

Visit the Taylor & Francis Web site at
http://www.taylorandfrancis.com

and the CRC Press Web site at
http://www.crcpress.com

Dedication

This book is dedicated to Professor Angelo Marcello Anile, who is not only the inspiration for this book, but, in many ways, its main author. Beyond his individual professional creativity, he has guided and mentored so many, and has always been willing to share his knowledge in the seeds he has planted among young and seasoned colleagues alike.

Contents

Preface

Surfaces are a major component of geographical information systems (GISs). When surfaces are used to model geographical entities, they inherently contain uncertainty in terms of both position and attribute. This book presents the requisite framework for surfaces where uncertainty is an explicit part of the process, from the data input stage to the conceptualization of its storage as fuzzy geographical entities, to its display and to its analyses.

The reader will find the pseudo code of the computer programs that are requisite for actualizing what is presented in the book, while the associated code itself may be run utilizing the accompanying CD. The applications that are presented illustrate the concepts and indicate the relevance, richness, breadth, and depth of analyses that are possible with fuzzy surfaces. We trust that the reader will find what is presented to be useful.

The Editor

I was born and raised in São Paulo, Brazil, where I lived until the end of high school. I came to the United States and attended Muskingum College in New Concord, Ohio, where I graduated in 1967 with a major in mathematics (honors), a minor in physics, and an emphasis in philosophy. I obtained my master's degree from the University of Cincinnati in 1969 and a Ph.D. in mathematics from Oregon State University. In between my master's and Ph.D., I taught junior high and high school mathematics in rural Georgia from 1969 to 1973. I left Oregon State University in 1977 to begin work at Michigan State University as a systems analyst on an international project analyzing food production potential working in the Dominican Republic, Costa Rica, Nicaragua, Honduras, and Jamaica. In addition, I helped to develop software for food production analysis for Syria. My job consisted of developing software, geographical information systems, statistical models, linear programming models, analysis, and training for transfer to various countries. While in Costa Rica, I worked directly with the Organization of American States (IICA) on Nicaraguan and Honduran projects that were similar to those I worked on while at Michigan State. In 1982, I was hired by the Department of Mathematics of the University of Colorado at Denver where I am a full professor of mathematics.

Weldon A. Lodwick

Contributors

Marcello Anile
University of Catania
Dipartimento di Matematica e
    Informatica
Città Universitaria
Catania, Italy

Cidália Fonte
Universidade de Coimbra
Departamento de Matematica
Coimbra, Portugal

Giovanni Galo
University of Catania
Dipartimento di Matematica e
    Informatica
Città Universitaria
Catania, Italy

Gil Gonçalves
Universidade de Coimbra
Departamento de Matematica
Coimbra, Portugal

Jorge Santos
Universidade de Coimbra
Departamento de Matematica
Coimbra, Portugal

Salvatore Spinella
University of Catania
Dipartimento di Matematica e
    Informatica
Città Universitaria
Catania, Italy

1 Introduction

Weldon Lodwick, Marcello Anile, and Salvatore Spinella

CONTENTS

Surfaces are central to geographical analysis and their generation and manipulation a key component of geographical information systems (GISs). Geographical surface data are often not precise (errorless) since uncertainty is inherent in what is obtained — measurement uncertainty (position, instrument accuracy), method of data collection (satellite, sonar, LIDAR, altimeter, air photographs), the surface itself (ocean bottom, steep slope, terrain, wetlands, landforms — categories whose meanings combine several definitions or ideas), and methods of classification (dense forest, land cover/use — classifications, for example, defined via clustering algorithms). The uncertainty types of interest to this presentation are: (i) finite **ranges** or **intervals**, for example, slope data are often given in 10-degree increments; (ii) **transitional boundaries** or **fuzzy sets**, for example, ocean to shore, grassland to shrubland; (iii) **possibilistic values** which are values that are known to exist, such as a 2000-meter isoline, but whose precise location is not since it is based on evidence arising from measurement and knowledge of the area, for example; and (iv) **frequency** or **probability**, for example, the daily temperature distribution over the last 100 years. It is emphasized that these four types of uncertainty are distinct from each other and must be handled correctly semantically and analytically. Data of types (i), (ii), (iii), and (iv) will be called in this study **uncertainty data**.

This book presents the theory, methods, algorithms, and applications that directly model and analyze surfaces that are derived from uncertainty data. The focus is on the uncertainty that arises from transitional boundaries (intervals or fuzzy sets) and possibility distributions. We do not deal with statistical methods and probabilistic uncertainties since these are well known. That is, the uncertainties that are studied herein and the ways to deal with these uncertainties within the framework of a GIS and geographical analysis of surfaces are limited to three types — intervals, fuzzy sets, and possibility distributions. In the sequel, these are clearly described; the associated mathematics, as it pertains to surfaces defined; the algorithms developed; and the applications illustrated. The book sets out the process to identify the uncertainty in the geographical entity being studied, obtain the associated data, model, and analyze the data, display, and interpret results all within the context of a GIS. The book's development takes the reader from the foundations of uncertainty relevant to the type of data encountered in surface analysis, its representation, storage and manipulation to analysis and display.

The thesis set forward here is that much, if not most, of the uncertainty in geographical surface analysis is typically interval, fuzzy, and/or possibilistic. Second, when these data do contain one of these three types of uncertainty, the most faithful, useful, and natural way to model surfaces is to incorporate (represent and store), analyze, and display explicitly using interval, fuzzy, and/or possibilistic entities. To this end, the theory, methods, and applications are presented that clearly demonstrate the way to do this.

A reader interested in understanding how to create, analyze, and display surfaces with inherent interval, fuzzy, or possibilistic uncertainties will find help herein to perform the task via the software and applications presented in the book. Moreover, the reader will develop a deeper knowledge of uncertainty itself as related to geographical analysis of surfaces containing interval, fuzzy, and possibilistic uncertainties so that the applications of these theories to problems that arise yield solutions that are more closely allied to the inherent reality, which is usually and most often undeniably imprecise and fuzzy. Since the three aspects of uncertainty that are the focus of this book (interval, fuzzy, and possibility) are relatively new fields of study, a few paragraphs introducing their development follow, much of which is a summary of [68].

Fuzzy set and possibility theory were defined and developed by L. Zadeh beginning with [103] and subsequently [104], [105], [106], and [107]. As is now well known, the idea was to mathematize and develop analytical tools to solve problems whose uncertainty was more ample in scope than probability theory. Classical mathematical sets, for example, a set A, have the property that either an element $x \in A$ or $x \notin A$, but not both. There are no other possibilities for classical sets, which are also called *crisp* sets. An interval is a classical set. Zadeh's idea was to relax this "all-or-nothing" membership in a set to allow for grades of belonging to a set. When grades of belonging are used, a fuzzy set ensues. To each fuzzy set $\tilde{A}$, Zadeh associated a real-valued function $\mu_{\tilde{A}}(x)$, called a *membership function*, for all x in the domain of interest, the universe Ω, whose range is in the interval $[0, 1]$ that describes and quantifies the degree to which x belongs to $\tilde{A}$. For example, if $\tilde{A}$ is the fuzzy set "warm day," then a 0°C day has a membership value of 0 while a 20°C day might have a membership value of 1, and a 10°C day might have a membership value of one half. That is, a fuzzy set is a set for which membership in the set is defined by its membership function $\mu_{\tilde{A}}(x) : \Omega \to [0, 1]$ where a value of 0 means that an element does not belong to the set $\tilde{A}$ with certainty and a value of 1 means that the element belongs to the set $\tilde{A}$ with certainty. Intermediate values indicate the degree to which an element belongs to the set. Using this definition, a classical (so-called crisp) set A is a set whose membership function has a range which is binary, that is, $\mu_A(x) : \Omega \to \{0, 1\}$, where $\mu_A(x) = 0$ means that $x \notin A$, and $\mu_A(x) = 1$ means $x \in A$. This membership function for a crisp set A is, of course, the characteristic function. So a fuzzy set can be thought of as being one which has a generalized characteristic function that admits values in $[0, 1]$ and not just two values $\{0, 1\}$ and is uniquely defined by its membership function. Another way of looking at a fuzzy set is as a set in R^2 as follows.

DEFINITION 1.1
A fuzzy set $\tilde{A}$, as a crisp set in R^2, is the set of ordered pairs

$$\tilde{A} = \{(x, \mu_{\tilde{A}}(x))\} \subseteq \{(-\infty, \infty) \times [0, 1]\}. \tag{1.1}$$

Much has been written about fuzzy sets that can be found in standard textbooks (see, for example, [62]) and will not be repeated here. We present only the ideas that are pertinent to the areas in the interfaces between interval and fuzzy analysis of interest.

The use of fuzzy sets in geographical analysis is relatively new (see [5], [41], [44], [46], [48], [66], [69], [71], [88], and [99]). Its application to geographical surfaces is even more recent ([5], [69], and [88]). Fuzzy surfaces will be studied in detail in subsequent chapters.

An interval may be represented as a fuzzy set and consequently as a possibility distribution, as we will see below. Given this fact, we will concentrate on fuzzy set and possibilistic uncertainties. However, it will be via intervals that we will develop our analytical and numerical techniques associated with fuzzy sets. Since interval numbers are easier to manipulate than fuzzy sets, intervals and their applications to surfaces will be developed separately.

Possibility theory was also developed by Zadeh [108] and extended by many authors, notably by Dubois and Prade (see [27], [32]), to model uncertainties to allow for a more general theory of uncertainty than probability theory models. There is often confusion in the semantics of uncertainty pertaining to probability, interval, fuzzy, and possibility. This is clarified below. GIS applications have seldom, if ever, used possibilistic geographical analysis. There are many reasons for this. Perhaps the greatest reason is that fuzzy set theory, as distinguished from possibility theory, is not always clear. Second, since geographical entities are often fuzzy (boundaries are gradual or transitional in nature between geographical entities) the use of possibilistic entities is frequently omitted. Third, since Zadeh develops possibility theory via fuzzy set theory [108], most authors do not make a distinction and consider possibility distributions the same as fuzzy membership functions. The distinction between fuzzy set theory and possibility theory is most important in its semantics so that a section dealing with the semantics is included below.

The first research papers in fuzzy set theory appeared in the early 1960s, and while fuzzy "logic" (which can be related to mathematical analysis) can be thought of being developed in the early 1920s, the area of fuzzy set theory as a separate field of study dates from the mid-1960s. On the other hand, the first research "paper" in interval analysis can be considered to be Archimedes' computation of circumference of a circle [83]. More recently Burkhill [13] in 1924 can be considered as the second research paper in the field. However, interval analysis as a separate field of study began in the early 1960s. While there are five known direct and clear precursors to Moore's version of interval arithmetic and interval analysis beginning in 1924 (see [13], [35], [91], [97], and [102]), Moore was the one who worked out rounded computer arithmetic and fully developed the mathematical analysis of intervals, called interval analysis. As developed by R. E. Moore (see [78], [81], and [82]), interval analysis arose from the attempt to compute error bounds of numerical solutions on a finite state machine that accounted for all numerical and truncation errors, including roundoff error, automatically (by the computer itself). This led in a natural way to the investigation of computations with intervals as the entity, data type, that enabled automatic error analysis.

R. E. Moore and his colleagues are responsible for developing the early theory, extensions, vision, and wide applications of interval analysis and the actual implementation of these ideas to computers. E. R. Hansen writes (see [53]):

> R. E. Moore (see [80]) states that he conceived of interval arithmetic and some of its ramifications in the spring of 1958. By January of 1959, he had published [78] a report on how interval arithmetic could be implemented on a computer. A 1959 report [82] showed that interval computations could bound the range of rational functions and integrals of rational functions. Theoretical and practical interval arithmetic were differentiated. Reference [77] discusses interval valued functions, interval contractions, a metric topology for interval numbers, interval integrals, and contains an extensive discussion of Moore's use of interval analysis to bound the solution of ordinary differential equations.

The methods for handling interval uncertainty and arithmetic of fuzzy sets, thus, can be traced to R. E. Moore's technical reports in 1959 and 1960, and his Ph.D. thesis in 1962 (see [78], [79], [81], and [82]). On the real number line, with the usual meaning of the order relation, an interval $[a, b]$ is the *set* of all real numbers $\{x | a \le x \le b\}$. Moreover, intervals can be considered as a number with two parameters (the left endpoint and the right endpoint). These two natures of intervals, *numbers* and *sets*, lead to interval arithmetic and interval analysis. What follows will develop the methods required to represent, store, manipulate, and analyze interval entities. While interval analysis is over 40 years old, the application of interval analysis to geographical surfaces is recent (see [66], [69], [71], and [88]).

Intersections and unions of intervals also have an algebra and are computed in a straightforward manner. These definitions, unlike those of fuzzy set theory found in the sequel, come from classical set theory. Intersections and unions, of course, are crucial in defining what we mean by solutions to simultaneous equations and inequalities as well as the fundamental building blocks of logical statements. For a given interval $[a, b]$ and a given real number x, the statement $x \in [a, b]$ is either true or false. There is no vagueness or ambiguity except for roundoff when the statement is implemented on a computer. For two intervals A_1 and A_2, if we know that $x \in A_1$ and $x \in A_2$, then we also know with certainty that $x \in A_1 \cap A_2$. These statements have certainty except when one accounts for implementations on the computer and roundoff error comes into play unlike statements of this type in fuzzy set theory. Interval arithmetic and the interval analysis developed from it do not assign any measure of possibility or probability to parts of an interval. A number x is either in an interval A, or it is not. By introducing probability distributions or possibility distributions on an interval, and using level sets, integrals, or other measures, a connection between intervals and fuzzy sets can be made.

Intervals are sets, and they are a (new type of) number. This dual role is exploited in the arithmetic and analysis. As a **fuzzy set**, an interval number $U = [\underline{u}, \bar{u}]$ has a membership function

$$\mu_U(x) = \begin{cases} 1 \text{ for } \underline{u} \leq x \leq \bar{u}, \\ 0 \text{ otherwise} \end{cases}$$

Thus, interval analysis can be considered as a subset of fuzzy set theory. As a **probability distribution**, an interval number can be considered as one of two probability density functions:

1. An interval may be considered to be the uniform distribution

$$p(x) = \begin{cases} 1/(\bar{u} - \underline{u}) & \text{for } \underline{u} \leq x \leq \bar{u}, \ \underline{u} < \bar{u} \\ 0 & \text{otherwise} \end{cases}$$

2. An interval may represent the fact that all we know is the support so that every distribution $p(x)$ with support $supp\{p(x)\} = [\underline{u}, \bar{u}]$ is in the interval.

The first distribution may be appropriate when an approximation to the distribution is needed in the presence of no information about how the uncertainty is distributed as a "best" guess. Most researchers, however, consider the second probabilistic interpretation the most faithful probabilistic meaning of an interval as an uncertainty entity.

There are excellent introductions to interval analysis beginning with R. E. Moore's book [77] (also see other texts listed in the bibliography). A more recent introduction can be found in [21] and downloaded from: http://www.eng.mu.edu/corlissg/PARA04/READ_ME.html. Moreover, there are introductions that can be downloaded from the interval analysis website http://www.cs.utep.edu/interval-comp. On the interval analysis website, a list of languages that support interval data types is given. See http://www.cs.utep.edu/interval-comp/intlang.html. In particular, a MATLAB™ system called INTLAB that supports the interval data types and interval linear algebraic analysis can be found and downloaded from: http://www.ti3.tu-harburg.de/~

rump/intlab/index.html. INTLAB was used to develop the fuzzy interval arithmetic systems and analyses associated with this book (also see [86]).

1.1   INTERVAL ARITHMETIC

1.1.1   Intervals and Operations Among Intervals

An *interval* $[a, b]$ is an ordered pair of real numbers a, b with $a < b$. In a computer, in order to obtain exact lower and upper bounds on interval arithmetic operations, it is necessary to use the conservative form representation, which consists in representing a with the **largest machine number** a_L such that $a_L \leq a$ and b with the **lowest machine number** b_R such that $b_R \geq b$ [54].

An interval $X = [a, b]$ is *positive* if $a \geq 0$, and strictly positive if $a > 0$, *negative* if $b \leq 0$, and *strictly negative* if $b < 0$.

Two intervals $X = [a, b]$, $Y = [c, d]$ are equal if $a = c$ and $b = d$. Among intervals, it is possible to define a *partial ordering*:

$$[a, b] < [c, d] \Leftrightarrow b < c \tag{1.2}$$

Arithmetic operations among intervals are defined as follows.

Let *op* be any arithmetic operation among reals, $op = +, -, *, /$. If X, Y are two intervals, *op* between the two intervals is defined as:

$$X \text{ op } Y = \{x \text{ op } y | x \in X, y \in Y\} \tag{1.3}$$

Explicitly:

$$[a, b] + [c, d] = [a + c, b + d] \tag{1.4}$$

$$[a, b] - [c, d] = [a - d, b - c] \tag{1.5}$$

$$[a, b] * [c, d] = [\min(ac, ad, bc, bd), \max(ac, ad, bc, bd)] \tag{1.6}$$

If the interval Y does not contain 0, the *inverse* $1/Y$ is defined by:

$$1/Y = [1/d, 1/c] \tag{1.7}$$

and the *division* of X by Y is:

$$X/Y = X * (1/Y) \tag{1.8}$$

Among the elementary operations, it is convenient to introduce the following:
Power, defined by the following rules [54]:

$$[a, b]^n = \begin{cases} [1, 1] & \text{if } n = 0 \\ [a^n, b^n] & \text{if } a \geq 0 \vee (a \leq 0 \leq b \wedge n \text{ is odd}) \\ [b^n, a^n] & \text{if } b \leq 0 \\ [0, \max(a^n, b^n)] & \text{if } a \leq 0 \leq b \wedge n \text{ is even} \end{cases} \tag{1.9}$$

For the *square root* and the *logarithm*, we have when $[a, b]$ is strictly positive [60],

$$\sqrt{[a, b]} = [\sqrt{a}, \sqrt{b}] \qquad (1.10)$$
$$\log([a, b]) = [\log(a), \log(b)] \qquad (1.11)$$

and for the *exponential*

$$\exp([a, b]) = [\exp(a), \exp(b)] \qquad (1.12)$$

First we notice that the commutative and distributive laws hold also for intervals $X = [a, b]$, $Y = [c, d]$, and $Z = [e, f]$.

$$X + Y = Y + X \qquad (1.13)$$
$$X * Y = Y * X \qquad (1.14)$$
$$X + (Y + Z) = (X + Y) + Z \qquad (1.15)$$
$$X * (Y * Z) = (X * Y) * Z \qquad (1.16)$$

Furthermore, defining $0 = [0, 0]$ and $1 = [1, 1]$, one has

$$0 + X = X + 0 = X \qquad (1.17)$$
$$1 * X = X * 1 = X \qquad (1.18)$$

Now we give some examples.

The distributive law between product and difference does not hold as shown by the following example:

$$[1, 2] * ([1, 2] - [1, 2]) = [-2, 2]$$

and

$$[1, 2] * [1, 2] - [1, 2] * [1, 2] = [-3, 3]$$

Also the difference between two intervals is not zero:

$$[2, 5] - [2, 5] = [-3, 3]$$

The ratio X/X is not 1 as shown by

$$[2, 5]/[2, 5] = [2/5, 5/2]$$

This causes what is called the *redundancy problem*:

$$\frac{X * Y}{Y} \neq X$$

as an example:

$$[1, 3] * [2, 4]/[2, 4] = [1/2, 6]$$

and the resulting interval $[1/2, 6]$ is much larger than $x = [1, 3]$.

Therefore, whenever in an expression a variable is repeated several times, the interval domain grows in an uncontrollable way, and furthermore, this makes it impossible to adopt the standard cancellation laws when performing algebraic calculations.

This *redundancy problem* implies that, for the general problem of estimating the final uncertainty of a complex calculation, it is not practical to utilize interval arithmetic because there is no general way to control the growth of the uncertainty interval at each step of the calculation.

There are ways to remedy this drawback [54] by taking appropriate precautions, such as automatically analyzing the individual steps of a complex calculation in order to detect redundancies and correct them. A rather interesting approach is that of *affine arithmetic*, see [90], which has been applied to several problems in computer graphics. Another relevant approach is that of *constrained interval arithmetic*, see [67]. However, in the interval literature, there seems to be no generally accepted methodology and accompanying efficient software for achieving this.

In this book, we shall not utilize this general form of interval arithmetic (i.e., breaking a calculation into a sequence of interval operations and then applying the interval arithmetic rules) in order to treat uncertainty in interesting environmental problems. Instead, we shall adopt a more *ad hoc* approach, in the sense that, for a large class of functions that are encountered in interesting environmental problems, we shall utilize special algorithms, which estimate the final uncertainty interval with sufficient accuracy and efficiency.

1.1.2 EXAMPLES: THE MALTHUS LAW

One of the simplest examples is population dynamics as described by the discrete form of Malthus law:

$$X(t + 1) = \Lambda X(t) \qquad (1.19)$$

with Λ the growth constant. This constant (as well as other parameters that appear in population dynamics models, in epidemiological studies, etc.) can be interpreted as a transition probability in a Markov process. Usually, the data are not sufficient to determine the parameters with sufficient accuracy, and therefore the **incorporation of the uncertainty of the constant in the model is an important aspect of the model building**.

Recent work on modeling and forecasting tuberculosis in the United States with Markov processes has represented the uncertainty in the parameters using triangular fuzzy numbers [23], and as a consequence, also the predictions are represented in terms of fuzzy numbers.

In this chapter, we shall perform uncertainty analysis for the discrete Malthus law in Equation (1.19) with interval analysis and then we shall compare the results with those obtained via the probabilistic simulation approach. The fuzzy number analysis (which is in a sense an extension of the interval analysis, as will be shown later) will be performed in the next chapter.

Let us assume that the initial value of X lies in the range between 3 and 6 and therefore can be represented by the interval

$$X(0) = [3, 6]$$

and that the growth constant Λ lies in the range between 0.95 and 1.05 and is represented by the interval

$$\Lambda = [0.95, 1.05]$$

What will be the uncertainty if the population be at the 20th iteration? The answer

$$X(20) = [1.0754, 15.9198]$$

can be obtained with the MATLAB program **imalthus.m**:

```
clear; x0=infsup(3,6); lambda=infsup(0.95,1.05); x=x0;
mm=20; for
i=1:mm,
    x=lambda*x;
end
infsup(x)
y=(lambda^mm)*x0;
infsup(y)
```

In the probabilistic simulation approach, $X(0)$ and Λ are treated as uniform random variables in the ranges [3, 6] and [0.95,1.25], respectively. A naive application of Monte Carlo simulation through the MATLAB program **rmalthus.m** yields the histogram of the simulated data, and from this, the empirical probability density is calculated.

```
clear; n=20; m=3000000;
for j=1:m,
    x(1,j)=3+3*rand;
end
hist(x,20);
figure;
mean=1.;
for j=1:m,
  for i=2:n,
    lambda=0.95+0.1*rand;
    x(i,j)=x(i-1,j)*lambda;
  end
end
xx=1:m;
for j=1:m,
 yy(j)=x(n,j);
end
min(yy)
max(yy)
hist(yy,100);
figure;
zz=1:n;
Nx=hist(yy,zz);
```

```
zn=max(Nx);
for j=1:n,
 Nx(j)=Nx(j)/zn;
end plot(zz,Nx);
z=0;
for j=1:m,
 z=z+x(n,j);
end
z=z/m
```

With $m = 100$ simulation, one finds that the range of the values after 20 iterations is:

$$X(20) = [2.7719, 7.5400]$$

and one can see that the range is smaller than that obtained with interval arithmetic. By increasing the number of simulations to $m = 1000$, one obtains:

$$X(20) = [2.1289, 7.4606]$$

and for 10,000 simulations:

$$X(20) = [2.1181, 8.8219]$$

Finally, for 1,000,000 simulations, one has the histogram in Figure 1.1:

$$X(20) = [1.7384, 9.2908]$$

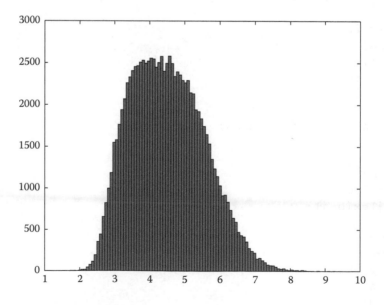

FIGURE 1.1 Histogram after 1,000,000 simulations.

which starts to approach the interval result. To approach more closely the interval arithmetic results, more Monte Carlo simulations are needed and, with the present naive approach, this would lead to incredible computing time.

Obviously, for the simple problem at hand, the probabilistic approach could be done in a much more efficient way by computing an integral. However, the scope of the exercise was to show the potential of the interval arithmetic approach over the probabilistic-simulation one, at least when the calculations could be done exactly (or almost exactly).

In this case, the interval obtained with interval calculations is exact because it expresses simply

$$X(n) = \lambda^n \cdot X(0) \tag{1.20}$$

and λ^n is calculated exactly according to the definition of power of an interval. The result is also consistent with the observation that the real function λ^n is monotonic. We notice that the largest interval obtained with the probabilistic approach requires 1,000,000 simulations, which is very demanding computationally. To obtain 1,000,000 function evaluations for functions that are not as simple as the Malthus law might be prohibitively expensive for the computational resources.

1.1.3 FUNCTIONS OF INTERVALS

It is convenient to introduce the subject with an example. Let us compute the average of two interval quantities:

$$Y = (C_1 * X_1 + C_2 * X_2)/(X_1 + X_2)$$

where

$$X_1 = [6, 8], \quad X_2 = [10, 12]$$

and the weights are also intervals:

$$C_1 = [600, 1500], \quad C_2 = [500, 1200].$$

Then using interval arithmetic, we obtain

$$Y = [430, 1650]$$

through the MATLAB program **average.m**

```
c1=infsup(600,1500);
c2=infsup(500,1200);
x1=infsup(6,8);
x2=infsup(10,12);
y1=(c1*x1+c2*x2)/(x1+x2);
y=infsup(y1)
```

This result is awkward because one expects that the average should be included in the interval [500, 1500].

In fact, a probabilistic simulation with $n = 10,000$ runs done with the MATLAB program **rmediau.m**

```
clear all;
n=50000;
x1l=6;
x1r=8;
c1l=600;
c1r=1500;
x2l=10;
x2r=12;
c2l=500;
c2r=1200;
x1sig=(x1r-x1l);
x2sig=(x2r-x2l);
c1sig=(c1r-c1l);
c2sig=(c2r-c2l);
for i=1:n,
    x1(i)=x1l+x1sig*rand;
end
for i=1:n,
    x2(i)=x2l+x2sig*rand;
end
for i=1:n,
    c1(i)=c1l+c1sig*rand;
end
for i=1:n,
    c2(i)=c2l+c2sig*rand;
end
for i=1:n,
    c(i)=(x1(i)*c1(i)+x2(i)*c2(i))/(x1(i)+x2(i));
end
xx1=mean(x1)
x1va=var(x1)
hist(x1,20)
figure;
xx2=mean(x2)
x2va=var(x2)
hist(x2,20)
figure;
cc1=mean(c1)
c1va=var(c1)
hist(c1,20)
figure;
cc2=mean(c2)
c2va=var(c2)
hist(c2,20)
figure;
cc=mean(c)
```

```
vc=var(c)
hist(c,20)
min(c)
max(c)
```

gives:
$$Y = [563, 1310]$$

which instead is well within the expected range [500,1500]. If we repeat the simulation with $n = 20,000$ runs we obtain

$$Y = [543, 1321]$$

Again, with $n = 30,000$ runs, we have

$$Y = [562, 1316]$$

and with $n = 40,000$ runs
$$Y = [536, 1326]$$

and the range seems to converge as it is shown by increasing n (however, at the cost of long CPU times).

The result obtained with a naive application of interval arithmetic is an example of the redundancy problem, i.e., a consequence of the repeated occurrence of the variables in the formula and indicates the need for a *different algorithm*. One of the most popular approaches is that of *interval weighted averages* also called the *vertex method* and this will be the subject of the next section.

1.1.4 THE VERTEX METHOD

First let us introduce functions of interval variables. Starting from a function of real variables one can introduce a function of interval variables:

DEFINITION 1.2
(Functions of interval variables) Let $f(x_1, \ldots, x_n)$ be a function of n real variables. Let $F(X_1, \ldots, X_n)$ be a function of n intervals $X_1, \ldots, X_n$. The function F is an interval extension of f if

$$f(x_1, \ldots, x_n) \in F(x_1, \ldots, x_n) \qquad (1.21)$$

whenever the intervals $X_1, \ldots, X_n$ reduce to the real numbers $x_1, \ldots, x_n$.

DEFINITION 1.3
(Inclusion monotone function) An interval function F is inclusion monotonic if

$$X_i \subset Y_i, \quad i = 1, \ldots, n \Rightarrow F(X_1, \ldots, X_n) \subset F(Y_1, \ldots, Y_n) \qquad (1.22)$$

The interval arithmetic operations are inclusion monotonic.

Now we can introduce the IWA algorithm of Dong and Wong [26] (also called the vertex method) in order to evaluate a large class of interval functions. Let the function to evaluate be

$$Y = f(X_1, \ldots, X_n) \tag{1.23}$$

with $X_1 = [a_1, b_1], \ldots, X_n = [a_n, b_n]$. We assume that f has **no extreme in the interior of the polyinterval** $[a_1, b_1] \times \ldots \times [a_n, b_n]$. Let us consider the $M = 2^n$ permutations of the endpoints, each represented by a vector with components β_i, $i = 1, \ldots, M = 2^n$:

$$\beta_1 = (a_1, a_2, \ldots, a_n)$$
$$\beta_2 = (b_1, a_2, \ldots, a_n)$$
$$\ldots$$
$$\beta_M = (b_1, b_2, \ldots, b_n)$$

Then the interval value of the function f is:

$$Y = [c, d] = [\min_i f(i), \max_i f(i)] \tag{1.24}$$

where $f(1) = f(a_1, a_2, \ldots, a_n)$, etc.

The MATLAB program **iwaff.m** performs such algorithm.

```
function y=iwaff(m,x,handle);
%implements the iwa algorithm with a general function;
%m is the number of arguments;
%x is the vector of arguments;
%to change the function one has to type first from the
%command window  handle=@function name;
for h=1:m;
    con(h)=0;
    t(h)=0;
end
for r=1:m,
    xl(r)=inf(x(r));
    xr(r)=sup(x(r));
end
for j=1:2^m,
    for i=1:m,
        if con(i)==2^(i-1);
            t(i)=mod(t(i)+1,2);
            con(i)=0;
        end
        xx(j,i)=(1-t(i))*xl(i)+t(i)*xr(i);
        con(i)=con(i)+1;
    end
end
for k=1:2^m,
```

```
    v=xx(k,:);
    ou(k)=feval(handle,v);
end
vv1=min(ou);
vv2=max(ou);
cc=infsup(vv1,vv2);
y=infsup(cc);
```

For the example we have considered, the result is the interval [533.33, 1333.33], which is the correct value. The **IWA** algorithm gives the correct result if the function has its extreme only on the boundary of the domain where it is defined. For the function under consideration in the example, this is the case, as can be verified by computing the derivatives and noticing that they do not vanish.

1.2   FUZZY ARITHMETIC

1.2.1   INTRODUCTION TO FUZZY SETS

1.2.1.1   Fuzzy Sets

Fuzzy sets represent a mathematical concept in order to describe mathematically nonstochastic uncertainty, for example, deriving from subjective judgments or from imprecise knowledge of parameters. A *fuzzy set* is a generalization of the classical concept of set. A fuzzy set A is an assembly of objects of the universe of discourse U that share some common property. The fuzzy set A is characterized by a membership function

$$\mu : U \rightarrow [0, 1] \qquad (1.25)$$

which associates to each element x of U a real number belonging to the interval [0, 1], representing the degree of belonging to A [62]. In this way, the concept of characteristic function of a set A is generalized. The concept of degree of belonging can be interpreted also in the framework of possibility theory as the degree of possibility that the event corresponding to x belongs to the class of events described by A [62]. The set of elements of U for which the membership function is positive is called the support of A. A singleton is a fuzzy set whose support consists of a single point of U.

In the following paragraphs, we shall give examples of membership function constructions from observed data.

1.2.1.2   Examples and Interpretations of the Membership Functions

Measurements in a Vague Environment: The Interpretation of Klawonn

Klawonn [61] introduces the membership function of a fuzzy set as a way of representing imprecision in a vague environment. It is better to start with an example.

Let us consider the temperature in a room in order to control it. For most purposes, a temperature difference less than 0.001°C is completely irrelevant for conditioning. Therefore, we can identify two temperature values that differ by less of 0.001°C.

More generally, once a tolerance threshold ϵ has been chosen (and this choice depends on the type of application), we can identify values of a variable whose distance (measured in an appropriate metric) s is less than ϵ.

As an example, consider the temperature in a room that is assumed to vary in the interval [0, 35] centigrades [61]. For the purpose of controlling the temperature, the region of most interest is that between 19 and 23°C, because human beings are rather sensitive to small variations within this range. Instead, the temperatures within the region 15–19 can be lumped in the concept of **cold** and that between 23 and 27 as **hot**, and from the viewpoint of the human sensations, it is not important to discriminate very accurately within those regions. Also, for temperatures in the ranges [0, 15] or [27, 35] there is no interest in fine-tuning the control because the only sensible choice would be either to heat up in the first case or to cool down in the second one.

In order to take this into account, it is convenient to distinguish the various scales by introducing a scale factor $c(x)$ and the operation defines a vague environment. Returning to the example of the temperature, a reasonable scale factor could be (see Figure 1.2):

$$c(x) = \begin{cases} 0 & \text{if } 0 \leq x < 15 \\ 0.25 & \text{if } 15 \leq x < 19 \\ 1.5 & \text{if } 19 \leq x < 23 \\ 0.25 & \text{if } 23 \leq x < 27 \\ 0 & \text{if } 27 \leq x < 35 \end{cases}$$

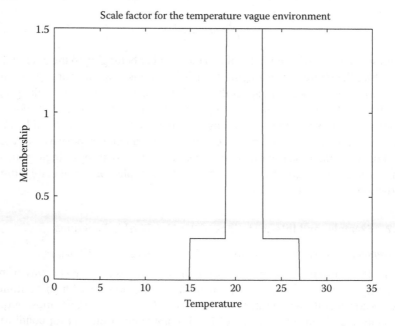

FIGURE 1.2 Scale factor for the temperature environment.

We shall consider two values x_1, x_2 ϵ-indistinguishable if their distance measured by the scale factor c (whose choice depends on the variability of x) is less than ϵ. Formally, we proceed as follows:

Let $X = [a, b]$ be an interval of the real line and $c(x)$, a scale factor (the only requirement on $c(x)$ is that it must be integrable). We can define a weighted distance between the two points x_1, x_2 belonging to the interval $[a, b]$

$$\delta_c(x_1, x_2) = \int_{x_1}^{x_2} c(s)\, ds \qquad (1.26)$$

The two points x_1, x_2 are considered ϵ-distinguishable with respect to the scale factor c if their transformed distance $\delta_c(x_1, x_2) \geq \epsilon$.

Now let us consider $x_0 \in X$. We associate to x_0 all other x that are not ϵ-distinguishable from x_0 (with respect to the scale factor c) and for ϵ varying between 0 and 1, that is, the set

$$S_{x_0, \epsilon} = \{x \in X : \delta_c(x, x_0) \leq \epsilon\} \qquad (1.27)$$

We let ϵ vary between 0 and 1; different values can be taken into account by a suitable redefinition of the scale factor. This set can be described through the map

$$\mu_{x_0} : X \to [0, 1] \qquad (1.28)$$

$$x \to 1 - \min(\delta_c(x, x_0), 1) \qquad (1.29)$$

whence

$$S_{x_0, \epsilon} = \{x \in X : \mu_{x_0}(x) \geq 1 - \epsilon\} \qquad (1.30)$$

We can interpret μ_{x_0} as the membership function of the fuzzy set of those values which are ϵ-indistinguishable from x_0; ϵ vary between 0 and 1. The interpretation of the membership function is then the following. The interval corresponding to the α-level α is the set of points that are indistinguishable from x_0 for $\epsilon = 1 - \alpha$. Example:

Let us assume $c(x)$ to be constant. Then $\delta_c(x_1, x_2) = c(x_2 - x_1)$, whence

$$\mu_{x_0}(x) = 1 - \min\{c(x - x_0), 1\}$$

is a linear function that assumes its maximum at x_0 and the sides of the triangle have slope c.

The MATLAB program **klawonnplot.m** found in the accompanying CD gives Figure 1.3. Another example is obtained with the scale factor

$$c(x) = \frac{(x - x_0)}{s^2} \exp\left(\frac{-(x - x_0)^2}{2s^2}\right)$$

which yields, with $s = 0.1$, Figure 1.4.

For the scale factor of 1.2, the membership functions corresponding to the nominal values **15, 19, 21, 23, 27 degree celsius** are given in Figures 1.5, 1.6, 1.7, 1.8, and 1.9.

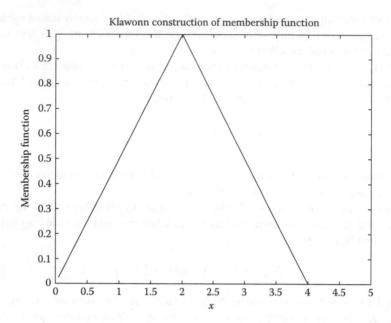

FIGURE 1.3 Membership function of almost 2 in a vague environment with a constant scale factor.

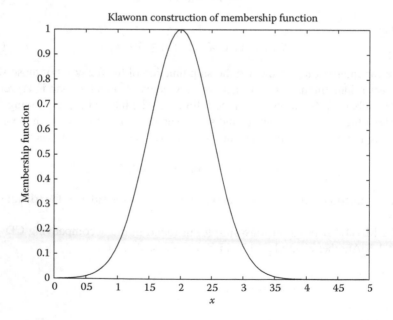

FIGURE 1.4 Membership function of almost 2 in a vague environment with a Gaussian scale factor.

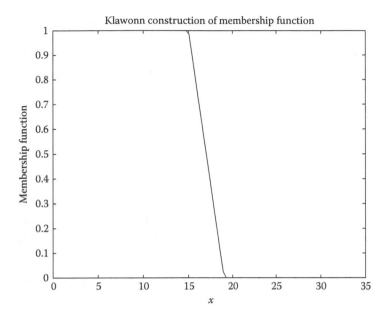

FIGURE 1.5 Membership function of 15 degrees in the vague environment defined by the temperature scale factor.

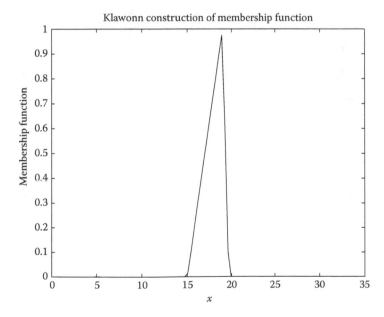

FIGURE 1.6 Membership function of 19 degrees in the vague environment defined by the temperature scale factor.

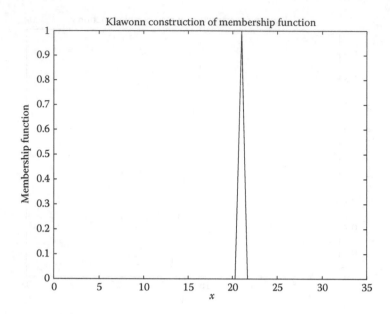

FIGURE 1.7 Membership function of 21 degrees in the vague environment defined by the temperature scale factor.

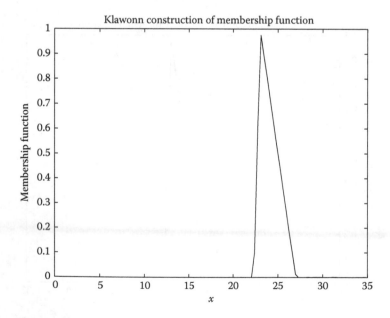

FIGURE 1.8 Membership function of 23 degrees in the vague environment defined by the temperature scale factor.

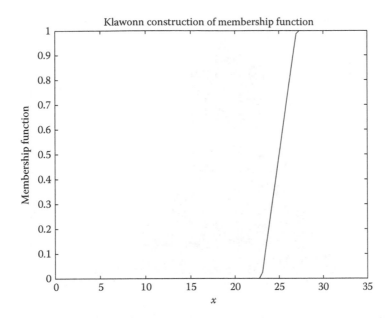

FIGURE 1.9 Membership function of 27 degrees in the vague environment defined by the temperature scale factor.

Membership Function as Degree of Preference for Design Parameters

According to the context, there could be several interpretations of the membership function. An interesting interpretation is due to Wood, Otto, and Antonsson [100]. In the design process, there could be several parameters, which could be imprecisely known, that is, they vary within some ranges. Therefore, a first step would be to treat them as intervals. However, the designer might have some preference for some of the parameters, and this could be quantified by introducing the degree of preference μ. This function could then be interpreted as the membership function of a fuzzy set.

Membership Function from Histograms: The Interpretation of Dubois and Prade

This method is better explained in the case of the definition of the fuzzy set concept of *tall*. The universe of discourse U consists of all the possible heights, for example, in centimeters. The interval $U = [l_{inf}, l_{sup}]$, where, for example, $l_{inf} = 0.5, l_{sup} = 2.5$ in meters.

A sample of people are asked at which height in the range $U = [l_{inf}, l_{sup}]$ they consider a person *tall*. If somebody answers, say, $s = 1.75$, this means that this person considers all individuals with height in the range $S = [s, l_{sup}]$ as tall. One can then discretize the set U in subsets of the kind $S_i = [s_i, l_{sup}]$, and the histogram of the numerical answers is given in Figure 1.10. The sets S_i form a nested sequence covering the set U. If the number of answers is considerably large, one can assign probabilities $P(S_i)$ to the sets S_i from the normalized histogram. Then a fuzzy set could

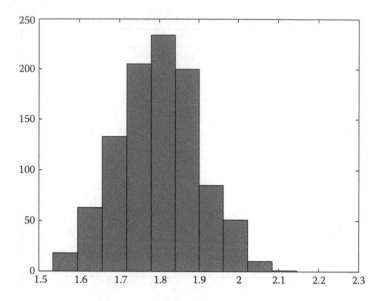

FIGURE 1.10 Histogram corresponding to the height distribution for the concept of *tall*.

be constructed by defining the membership function $\mu(s)$ on the set U as follows:

$$\mu(s) = \sum P(S_i) \qquad (1.31)$$

over $s_i \leq s$. The interpretation of this choice is that we assign to s the degree of belonging to the fuzzy set *tall* given by the sum of all probabilities that s lies in the intervals $[s_i, l_{sup}]$, which coincides with the cumulative distribution function of the probability measure defined by $P(S_i)$. This interpretation is quite intuitive and a formal justification in the framework of evidence theory can also be given [28].

With the **MATLAB** programs **DP1, DP1a, and DP1b** found in the accompanying CD, one obtains the graph of the membership function in Figure 1.11.

The Probabilistic Interpretation of Jamison and Lodwick

For a critical investigation of the relationship between membership function and probability, see Hisdal and Jamison and Lodwick [58].

1.2.1.3 The Mountain Function

The mountain function method aims at summarizing a set of data described with a single fuzzy number. Let the data set consist of the points:

$$DS = \{z_1, \ldots, z_N\} \qquad (1.32)$$

then the mountain function is defined by

$$M(c) = \sum_{j=1}^{N} exp(-\alpha d(z_j, c)) \qquad (1.33)$$

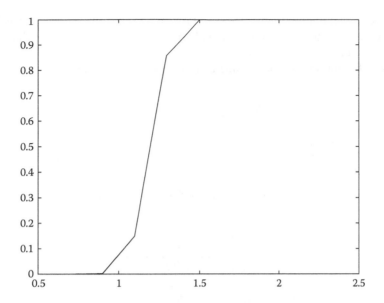

FIGURE 1.11 Membership function for the concept of *tall*.

where N is the number of data points, α is a positive constant, and $d(x_j, c)$ a distance, for example, of the kind

$$d(x_j, c_i) = (x_j - c)^2 \tag{1.34}$$

and c represents the modal point of the resulting fuzzy number. Then c is chosen as to maximizing the mountain function $M(c)$.

The method has been extended and modified in order to treat the case of imprecise data, described by triangular fuzzy numbers. For the sake of concreteness, let us represent the data as a set of quintuples of the kind:

$$D_j = (x_j, y_j, z_{jmin}, z_{jmax}, z_{jcenter}) \quad z_{jmin} < z_{jcenter} < z_{jmax} \tag{1.35}$$

For each quintuple, x_i and y_i represent the spatial coordinates, assumed to be precisely known, whereas $[z_{jmin}, z_{jmax}, z_{jcenter}]$ represents a fuzzy triangular number. In the particular case, when the three values of $[z_{jmin}, z_{jmax}, z_{jcenter}]$ coincide, we have a crisp number.

Our aim is to construct a fuzzy number that, in a sense, is a summary of the N fuzzy triangular numbers $[z_{jmin}, z_{jmax}, z_{jcenter}]$ located at (x_j, y_j). The construction proceeds as follows: Find a fuzzy number S with membership function m_S such that:

   i support$(m_S) = [a, b]$
     where $a = \min z_{jmin}$, $b = \max z_{jmax}$
  ii there exists a unique element z_S such that $m_S(z_S) = 1$
 iii m_S is monotone increasing to the left of z_S and decreasing to the right of z_S

In the case of imprecise data, we must take into account the spread of each datum in the definition of the mountain function. In particular, following [48] we shall weigh less the data that have a larger spread.

The definition we use for the mountain function is the following:

$$M(c) = \sum_{j=1}^{N} \exp(-d(z_j, c)) \tag{1.36}$$

where now the distance is given by

$$d(z_j, c) = (c - z_{jcenter})^2 \alpha_{j1} \quad c < z_{jcenter} \tag{1.37}$$

$$d(z_j, c) = (c - z_{jcenter})^2 \alpha_{j2} \quad c \geq z_{jcenter} \tag{1.38}$$

with

$$\alpha_{j1} = \frac{(z_{jcenter} - z_{jmin})^2}{2(z_{jcenter} - a)^2(z_{jmax} - z_{jmin})^2} \tag{1.39}$$

$$\alpha_{j2} = \frac{(z_{jcenter} - z_{jmin})^2}{2(z_{jcenter} - b)^2(z_{jmax} - z_{jmin})^2} \tag{1.40}$$

The modal point of the resulting fuzzy number is identified with the maximum of the mountain function. To construct the full membership function, one proceeds by utilizing a six–control point linear spline as follows: The control points P_0 and P_5 are, respectively, the points (a, 0) and (b, 0), implying that the support of the membership function is the interval [a, b]. The control point P_3 is chosen in such a way that the requirement **ii** is satisfied. The other control points P_1, P_2, and P_4 are chosen for the sake of simplicity, according to the following rules: If $(a - c) > (z - b)$, P_1 is taken to be:

$$P_1 = \left(c - \frac{3}{4}(b - c), h\right) \tag{1.41}$$

$$h = \frac{9(b - c)^2 k}{2(3c - 2a - b)(5b - 4a - c)}$$

$$k = \frac{(b - c)^2}{2(c - a)^2}$$

P_2 is taken to be:

$$\left(c - \frac{1}{2}(b - c), k\right) \tag{1.42}$$

P_4 is taken to be:

$$\left(c + \frac{1}{2}(b - c), k\right) \tag{1.43}$$

Similar formulae apply in the case $(a - c) \leq (z - b)$.

As a simple example, let us consider the following artificial data set:

$$D1 = (x1, y1, 109, 125, 144); \tag{1.44}$$

$$D2 = (x2, y2, 102, 125, 149); \tag{1.45}$$

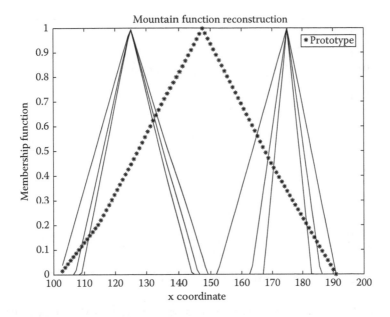

FIGURE 1.12 Mountain function reconstruction of a set of fuzzy data.

$$D3 = (x3, y3, 107, 125, 146);\qquad(1.46)$$
$$D4 = (x4, y4, 152, 175, 186);\qquad(1.47)$$
$$D5 = (x5, y5, 163, 175, 191);\qquad(1.48)$$
$$D6 = (x6, y6, 167, 175, 183).\qquad(1.49)$$

The spatial coordinates of the data are irrelevant for our purpose, because they are assumed to belong to the same cell. The disc that accompanies this book contains the **MATLAB** programs **distm.m, prototype.m,mountain.m**, which lead to the membership function represented in Figure 1.12.

1.2.1.4   The Median Construction of the Membership Function

An overly simplistic method for constructing a membership function from a set of data is that of fitting a triangular membership function to the reduced data set consisting of the minimum, the maximum, and the median. The median is taken to correspond to the peak value for $\alpha = 1$, and the minimum and maximum define the range of the triangular fuzzy number.

A better interval-based construction can be envisaged in which the support of the membership function is defined by the interval of the minimum and maximum of the data and the interval corresponding to the α-level α is obtained by deleting from the ordered data set, in a symmetric way from the median, a fraction α of the data from the range starting from the extremes. As an example consider the data consisting of the hourly measurements of the pollutant $PM10$ (dust) at a given site

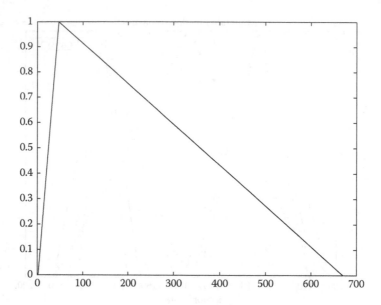

FIGURE 1.13 Triangular fuzzy number from a set of pollutant PM10 concentration data during the first 2 months of 1999 at a site near a petrochemical factory in Sicily.

near a petrochemical factory in Sicily, taken for the first 2 months of 1999. First we summarize the data with a triangular fuzzy number and obtain Figure 1.13. Next we apply the median method and obtain the much more satisfactory membership function of Figure 1.14. We consider also the yearly data from another station and obtain the triangular fuzzy number (Figure 1.15) and the median reconstruction (Figure 1.16). In this latter example, one sees clearly that the median reconstruction gives results that are much more satisfctory than the triangular fuzzy number representation. In particular, the effect of outliers is insignificant in the median reconstruction.

1.2.1.5   Operations with Fuzzy Sets

DEFINITION 1.4
(Union) Given two fuzzy subsets A, B of the universe of discourse U, defined by the membership functions μ_A and μ_B, their union is defined as the fuzzy subset of U whose membership function is

$$\mu_{A \cup B}(x) = \max(\mu_A(x), \mu_B(x)) \tag{1.50}$$

DEFINITION 1.5
(Intersection) Likewise the intersection of the two fuzzy subsets A, B is defined by the membership function

$$\mu_{A \cap B}(x) = \min(\mu_{A(x)}, \mu_{B(x)}) \tag{1.51}$$

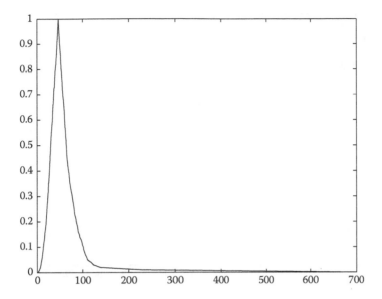

FIGURE 1.14 Median membership function construction from a set of pollutant PM10 concentration data during the first 2 months of 1999 at a site near a petrochemical factory in Sicily.

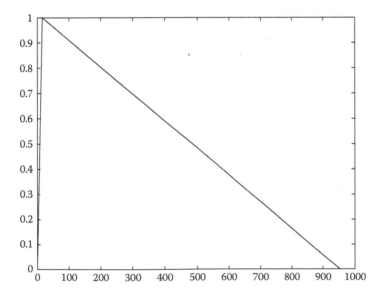

FIGURE 1.15 Triangular fuzzy number from a set of pollutant PM10 concentration data during the year 1999 at a site near a petrochemical factory in Sicily.

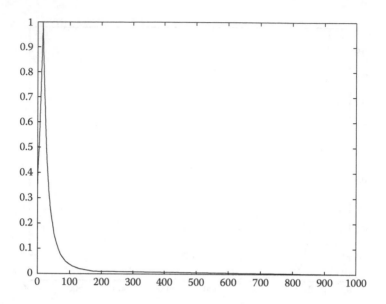

FIGURE 1.16 Median membership function construction from a set of pollutant PM10 concentration data during the year 1999 at a site near a petrochemical factory in Sicily.

DEFINITION 1.6
(Complement) The complement of the fuzzy subset A is defined by the membership function

$$\mu_{CA}(x) = 1 - \mu_{A(x)} \tag{1.52}$$

DEFINITION 1.7
(Equivalence) Two fuzzy subsets A and B are equivalent if their membership functions are equal:

$$A = B \Leftrightarrow \forall x \in U \ \mu_A(x) = \mu_B(x) \tag{1.53}$$

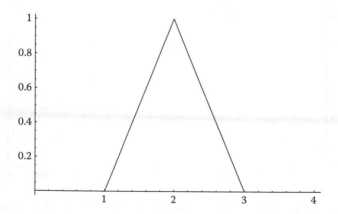

FIGURE 1.17 Triangular fuzzy number for $l = 1, m = 2, r = 3$ is **"almost 2."**

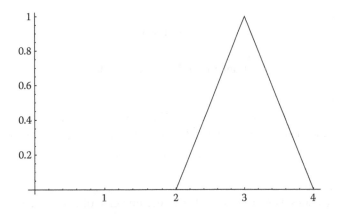

FIGURE 1.18 Triangular fuzzy number for $l = 2, m = 3, r = 4$ is **"almost 3."**

DEFINITION 1.8
(Inclusion) The fuzzy subset A is included in the fuzzy subsct B iff

$$A \subseteq B \Leftrightarrow \forall x \in U \ \mu_A(x) \leq \mu_B(x) \tag{1.54}$$

REMARK 1.1
(Identities) It is easy to prove the following identities:

$$C(C A) = A \tag{1.55}$$

$$C(A \cup B) = C A \cap C B \tag{1.56}$$

$$C(A \cap B) = C A \cup C B \tag{1.57}$$

In fact one has:

$$\mu_{CCA}(x) = 1 - \mu_{CA}(x)$$
$$= 1 - (1 - \mu_A(x))$$
$$= \mu_A(x)$$

and

$$\mu_{C(A \cup B)}(x) = 1 - \mu_{A \cup B}(x) \tag{1.58}$$
$$= 1 - max(\mu_A(x), \mu_B(x)) \tag{1.59}$$
$$= min(1 - \mu_A(x), 1 - \mu_B(x)) = \mu_{CA \cap CB}(x) \tag{1.60}$$

analogously for the other identity.

REMARK 1.2
(Tertium non datur) The law of "tertium non datur" does not hold; in fact

$$A \cup CA \neq U \tag{1.61}$$

because

$$max(\mu_A(x), 1 - \mu_A(x)) \neq 1$$

and likewise

$$A \cap CA \neq \emptyset \tag{1.62}$$

because

$$\min(\mu_A(x), 1 - \mu_A(x)) \neq 0$$

DEFINITION 1.9
(Fuzzy normal subset) A fuzzy subset A of U (universe of discourse) is said to be
normal if its membership function reaches its maximum at a point, that is,

$$\max_{x \in U} \mu_A(x) = 1 \tag{1.63}$$

DEFINITION 1.10
(Fuzzy convex subset) A fuzzy subset A of the universe of discourse U is said to be
convex if

$$\forall x, y \in [a, b] \quad \mu_A(\lambda x + (1 - \lambda)y) \geq \min(\mu_A(x), \mu_A(y)) \tag{1.64}$$

DEFINITION 1.11
(Direct product of fuzzy sets) Let U, V be universes of discourse and A, B fuzzy
subsets of U, V, respectively. The direct product $A \times B$ is the fuzzy subset of the
Cartesian product $U \times V$, defined by the membership function

$$\mu_{A \times B}(x, y) = \min(\mu_A(x), \mu_B(y)) \tag{1.65}$$

1.2.2 OPERATIONS AMONG FUZZY NUMBERS

1.2.2.1 Definition of Fuzzy Numbers

DEFINITION 1.12
(Fuzzy number) A fuzzy number is a convex and normal fuzzy with compact support
and piecewise continuous membership function.

The set of fuzzy numbers on the real line R will be denoted by $F(R)$.

DEFINITION 1.13
(α-levels) For a fuzzy subset A of the universe of discourse U, we introduce the
α-levels A_α defined as follows for $0 \leq \alpha \leq 1$

$$A_\alpha = \{x : \mu_A(x) \geq \alpha\} \tag{1.66}$$

PROPOSITION 1.1 (**Resolution principle**)
Let $\chi_{A_\alpha}(x)$ be the characteristic function of the α-level A_α, then

$$\mu_A(x) = \sup_{\alpha \in [0,1]} \min(\alpha, \chi_{A_\alpha}(x)) \tag{1.67}$$

An example of α-level representation of the membership function with MATLAB
is the m-file **atrian.m** included in the CD that accompanies this book.

As an example, we plot the triangular fuzzy number with l $= 1$, m $= 2$, r $= 3$, which could be interpreted as **almost 2**, Figure 1.17.

1.2.3 COMPARING FUZZY NUMBERS

The problem of comparing fuzzy numbers has been amply studied in the literature [29]. The simplest way of comparing two fuzzy numbers is to **defuzzify** the two numbers, obtaining two **crisp** numbers representing in a sense typical values of the two quantities, and then to compare these two crisp numbers in the usual way.

1.2.3.1 The Center of Mass

The simplest defuzzification procedure is that of taking the center of mass [111]:

$$g = \frac{\int_a^b x\mu(x)\,dx}{\int_a^b \mu(x)\,dx} \tag{1.68}$$

where $\mu(x)$ is the fuzzy number membership function.

The program MATLAB **centermass.m** found in the accompanying CD computes the center of mass for a fuzzy number given by the matrix of its α-levels.

1.2.3.2 The Mean Interval

A very interesting concept is that of **mean interval**, which in a sense summarizes both the expected value and the spread of a fuzzy number [29]. Its definition for a fuzzy number described by the α-*levels* $z_\alpha = [zl_\alpha, zr_\alpha]$ is

$$M(z) = \left[\int_0^1 zl_\alpha d\alpha, \int_0^1 zr_\alpha d\alpha \right] \tag{1.69}$$

The program MATLAB **meanint.m** found in the accompanying CD computes the mean interval for a fuzzy number given by the matrix of its α-levels.

1.2.3.3 Distances Between Fuzzy Numbers

A simple way of constructing distances for fuzzy numbers is to use any of defuzzifying scalars and defining a distance between two fuzzy numbers u, v in terms of the absolute distance between the corresponding defuzzifying scalars, for example:

$$dist_1(u, v) = Abs[MM[u] - MM[v]] \tag{1.70}$$

where $MM[u]$ represents the scalar that defuzzifies u.

Other distances could be defined on fuzzy numbers starting from distances on intervals.

1.2.3.4 Distance Between Intervals

A classical example of a distance between intervals is the Hausdorff distance

$$d_H(A, B) = \max[\sup_{x \in A} \inf_{y \in B} |x - y|, \sup_{y \in B} \inf_{x \in A} |y - x|] \quad \text{(Hausdorff)} \tag{1.71}$$

Another widely used distance [29] starts from a function $\phi_p(x) = x^p$ and two intervals $A = [a, b]$ and $B = [c, d]$. The distance is then:

$$dist(A, B) = \phi_p^{-1}(0.5(\phi_p(|\, a - c\,|) + \phi_p(|\, b - d\,|))) \qquad (1.72)$$

where $p \geq 1$.

Then given two fuzzy numbers in terms of their α-levels u_α, v_α, an induced distance can be defined by [29]:

$$dist_2(u, v) = \int_0^1 dist(u_\alpha, v_\alpha)\, d\alpha \qquad (1.73)$$

The program MATLAB **fdist.m** found in the accompanying CD computes the distance $dist_2$ for two fuzzy numbers given by the matrices of their α-levels.

However, both distances introduced above suffer from the drawbacks that they do not enjoy some desirable properties of distance between intervals, as discussed by Bertoluzza et al. [10]. To remedy these drawbacks arising from the fact that only the endpoints of the intervals are taken into account in the interval distance definition, a new class of distances with interesting properties has been introduced in [10] as follows.

DEFINITION 1.14
(Distance between intervals) Let g be a normalized and integrable weight function on $[0, 1]$. The squared distance d^2 between two intervals $A = [a, b]$ and $B = [c, d]$ is given by

$$d^2(A, B) = \int_0^1 [t \cdot \|a - b\| + (1 - t)\|c - d\|]^2 \, dg(t) \qquad (1.74)$$

In this class, a particular distance that corresponds to concentrated weights at the endpoints and at the midpoints of the intervals and is easily implemented is [10]:

$$d^2(A, B) = k(a - b)^2 + h(A_G - B_G)^2 + k(c - d)^2$$

with $A_G = a + b/2$, $B_G = c + d/2$, and the weights h, k are such that $2k + h = 1$.

Another distance based on similar concepts has been introduced by Tran and Duckstein ([72] and [94]) and applied to fuzzy regression:

$$d^2(A, B) = \int_{-\frac{1}{2}}^{\frac{1}{2}} \left(\left[\frac{1}{2}(a + b) + t(b - a)\right] - \left[\frac{1}{2}(c + d) + t(d - c)\right] \right)^2 dt$$

$$= \left[\frac{1}{2}(a + b) - \frac{1}{2}(c + d)\right]^2 + \frac{1}{3}\left[\frac{1}{2}(b - a) - \frac{1}{2}(c - d)\right]^2 \qquad (1.74a)$$

As an example, the distance **dist₂** between the intervals $A = [0, 0]$ (a crisp number) and $B = [-1, 3]$ is the same as between A and $C = [1, 3]$ and equal to

2.2361, although A is inside B and outside C. Instead, with the distance introduced by *Tran and Duckstein* [72], the distance between A and B is 1.5275 and that between A and C is 2.0817.

By utilizing either of the previously defined distances between intervals, one can define distances between fuzzy numbers as follows.

DEFINITION 1.15
(Distance between fuzzy numbers) Given an integrable weight function w, $[0, 1] \rightarrow R$ such that:

$$w(\alpha) \geq 0$$
$$\alpha' < \alpha'' \Longrightarrow w(\alpha') < w(\alpha'')$$
$$\int_0^1 w(\alpha)\, d\alpha = 1 \qquad (1.75)$$

the distance between fuzzy numbers $\widetilde{A}, \widetilde{B} \in F(R)$ is a function $d : F(R)^2 \rightarrow R$ *defined by*

$$d(\widetilde{A}, \widetilde{B}) = \sqrt{\int_0^1 d^2([\widetilde{A}]_\alpha, [\widetilde{B}]_\alpha) w(\alpha)\, d\alpha} \qquad (1.76)$$

This definition is consistent with the fact that intervals with a higher presumption level α should have a higher weight in determining the distance. Furthermore, this definition reduces to the Euclidean one for real numbers.

The MATLAB programs **fdistbert.m** and **fdistduck** found in the accompanying CD compute the above defined distances for two fuzzy numbers given by the matrices of their α-levels in terms of the interval distances given by Equations (1.75) and (1.76), respectively.

As an example let us consider the two couples of triangular fuzzy numbers:
$A = (-2, 0, 2)$, $B = (-1, 0, 1)$, $A_1 = (-2, 0, 1)$, $B_1 = (-1, 0, 2)$. The two couples are represented in Figures 1.19 and 1.20, respectively. From the figures, it is intuitive that the distance between A_1 and B_1 should be greater than that between A and B because in A_1 and B_1 there are points that are farther away than between A and B. Now with the first distance we have introduced in Equation (1.73), one finds:

$$fdist(A, B) = 0.5$$
$$\qquad (1.77)$$
$$fdist(A_1, B_1) = 0.5$$

which shows that such a distance is not able to discriminate between the two cases. Instead, for the other distances we have introduced:

$$fdistbert(A, B) = 0.4472$$
$$\qquad (1.78)$$
$$fdistbert(A_1, B_1) = 0.5774$$

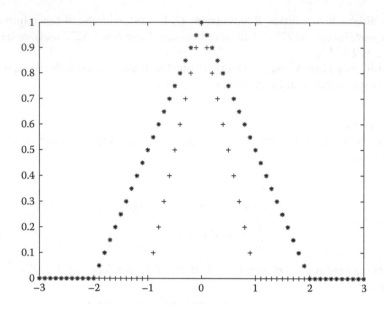

FIGURE 1.19 Triangular fuzzy numbers (−2, 0, 2) and (−1, 0, 1).

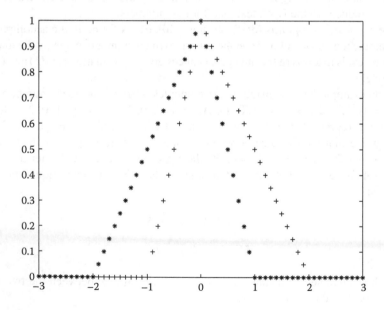

FIGURE 1.20 Triangular fuzzy numbers (−2, 0, 1) and (−1, 0, 2).

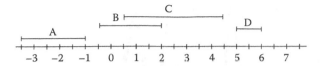

FIGURE 1.21 Interval A overtakes B by 0, B overtakes C by 3/5 and finally D overtakes C by 1.

and also

$$fdistduck(A, B) = 0.333$$
$$fdistduck(A_1, B_1) = 0.5774$$
(1.79)

which shows that these distances are more sensitive to the shapes of the fuzzy numbers.

With the introduced distance, one can obtain a consistent ordering for fuzzy numbers that is adequate for most practical applications. Given any distance between fuzzy numbers, we can give the following definition of equivalence between fuzzy numbers.

DEFINITION 1.16
(Fuzzy equivalence) Given two fuzzy numbers $\widetilde{A}$ and $\widetilde{B}$, a real number $\varepsilon > 0$ and a fuzzy distance $d : F(R)^2 \rightarrow R$, then $\widetilde{A}$ and $\widetilde{B}$ are ε-equivalent if $d(\widetilde{A}, \widetilde{B}) < \varepsilon$, that is,

$$\widetilde{A} -_{\varepsilon} \widetilde{B} \Longleftrightarrow d(\widetilde{A}, \widetilde{B}) < \varepsilon$$
(1.80)

In order to represent the inequality relationship, it is convenient to introduce the definition of *overtaking between fuzzy numbers*. As usual, we start with *overtaking between intervals*.

DEFINITION 1.17
(Overtaking between intervals) The overtaking of interval A with respect to interval B is the real function $\sigma : I(R)^2 \rightarrow R$ defined as:

$$\sigma(A, B) = \begin{cases} 0 & A^u \leq B^l \\ \frac{A^u - B^l}{width(A)} & A^u > B^l \wedge A^l \leq B^l \\ 1 & A^l > B^l \end{cases}$$
(1.81)

where $width(A)$ is the width of interval A.

Figure 1.22 clarifies the above definition. The overtaking of A with respect to B is 0, and of B with respect to C is 3/5 while of D with respect to C is 1. The program MATLAB **intover.m** included in the CD that accompanies this book computes the above defined overtaking between two intervals.

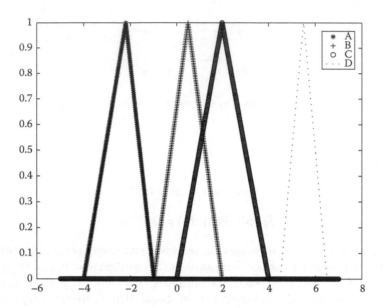

FIGURE 1.22 Triangular fuzzy numbers $(-4, -2.2, -1)$, $(-1, 0.5, 2)$, $(0, 2, 4)$, and $(4.5, 5.5, 6.5)$.

From this definition, one can define the δ-overtaking operator as:

DEFINITION 1.18
(δ-overtaking operator between intervals) Given two intervals A, B and a real number $\delta \in [0, 1]$, then A overtakes B by δ if $\sigma(A, B) \geq \delta$, that is,

$$A \geq_\delta B \iff \sigma(A, B) \geq \delta \qquad (1.82)$$

Likewise we can define the *overtaking between fuzzy numbers* as follows:

DEFINITION 1.19
(Overtaking between fuzzy numbers) One defines overtaking of the fuzzy number $\widetilde{A}$ with respect to the fuzzy number $\widetilde{B}$ as the real function $\sigma : F(R)^2 \to R$ defined as

$$\sigma(\widetilde{A}, \widetilde{B}) = \int_0^1 \sigma([\widetilde{A}]_\alpha, [\widetilde{B}]_\alpha) w(\alpha) \, d\alpha \qquad (1.83)$$

where for $w : [0, 1] \to R$, one assumes the definitions in Equation (1.75).

Figure 1.21 clarifies the above definition. The overtaking of A with respect to B is 0, and of B with respect to C is about 0.2430 while of D with respect to C is 1. The program MATLAB **over.m** included in the CD that accompanies this book computes the above-defined overtaking for two fuzzy numbers given by the matrices of their α-levels.

Likewise one can define the operator of δ-overtaking between fuzzy numbers as follows:

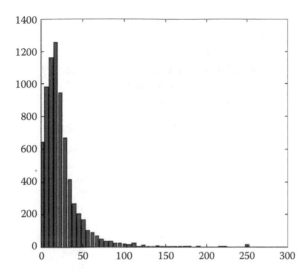

FIGURE 1.23 Histogram of the hourly PM10 data detected by a sensor for a whole year.

DEFINITION 1.20
(δ-overtaking operator between fuzzy numbers) Given $\widetilde{A}, \widetilde{B} \in F(R)$ and a real number $\delta \in [0, 1]$, then $\widetilde{A}$ overtakes $\widetilde{B}$ by δ if $\sigma(\widetilde{A}, \widetilde{B}) \geq \delta$, that is,

$$\widetilde{A} \succeq_\delta \widetilde{B} \longleftrightarrow \sigma(\widetilde{A}, \widetilde{B}) \geq \delta \qquad (1.84)$$

1.2.3.5 Comparison of Statistical and Fuzzy Descriptions

A comparison with a statistical approach has been made as follows: We consider the yearly pollution data previously introduced (see Figure 1.16). From an analysis of the data histogram, it is seen that they are reasonably well represented by a $Gamma(a, b)$ distribution (Figure 1.23).

Having computed the mean and standard deviation of the data (see Table 1.1), one has proceeded to fit such a distribution to a $Gamma(a, b)$ distribution.

A possible way to compare the fuzzy number membership function with the statistical distribution is to calculate the probability of observing values of the distribution greater than t and the overtaking of the fuzzy number representing the data with respect to the threshold t of a typical quantity, for example, $\delta = 0.66$. Table 1.2 suggests a possible comparison.

TABLE 1.1
Statistical Parameters Obtained from the Data Detected by the Environmental Sensor

No. Elements	Average	Standard Deviation	Minimum	Median	Maximum
7314	25.10	29.63	0.1	18.3	954.2

TABLE 1.2
Relationship between Thresholds and
Probability for the Chosen Distribution
Model and Comparison with the
Fuzzy-Overtaking Method

Threshold t	$P(X \geq t)$	$\widetilde{Z} \geq_{0.66} t$
1.27	0.9	True
3.45	0.8	True
6.37	0.7	True
10.10	0.6	True
14.82	0.5	True
20.93	0.4	False
29.16	0.3	False
41.22	0.2	False
62.64	0.1	False

The third column in the table considers the comparison between fuzzy number representing the data $\widetilde{Z}$ overtaking the threshold t by $\delta = 0.66$ and the probability of the random variable with the Γ distribution function to be greater than t. Notice the precise correspondence between probabilities and truth values. We are confident therefore that the rough membership function construction matches reasonably the statistical distribution function. However, a detailed statistical data analysis is required in order to guess the distribution function whereas the fuzzy number membership construction does not require any prior data analysis.

We remark that the *mean interval* of the median-reconstructed fuzzy number is [5.8719, 40.7382].

1.2.3.6 Arithmetic Operations. Examples

Let us consider the function of two variables $g : X \times Y \to Z$ and let A, B be fuzzy subsets of X, Y, respectively. Then, on the basis of the **extension principle**, one has

$$\mu_{g(A,B)}(z) = \sup[\mu_A(x), \mu_B(y)] \qquad (1.85)$$

where $z = g(x, y)$ and in terms of α-*levels*, $g(A, B)_\alpha = g(A_\alpha, B_\alpha)$. Therefore, the result is the same as by applying the interval arithmetic extension of the function g.

The MATLAB m-files **sum.m, product.m, difference.m, and quotient.m** given found in the CD that accompanies this book perform the four basic operations.

As an example, we consider the operations with triangular fuzzy numbers. Let $x = atrian(1, 2, 3)$ and $y = atrian(2, 3, 4)$ be two triangular fuzzy numbers (in this case, they are centered at $m = 2$ and $m = 3$, respectively, and could be interpreted as **"almost 2"** and **"almost 3"**) given in Figures 1.17 and 1.18. Their sum is then given by the fuzzy number $z = x + y$ in terms of α-*levels*, represented in Figure 1.24.

Likewise we have for the product of the previous two fuzzy numbers $z = x * y$, represented in Figure 1.25.

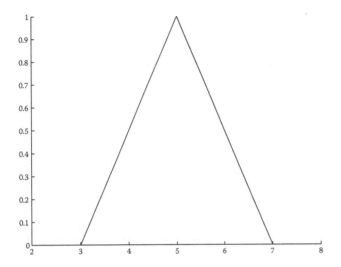

FIGURE 1.24 Sum of the TFN (triangular fuzzy numbers) $[1, 2, 3]$ and $[2, 3, 4]$.

For the inverse operations, we have the difference $z = y - x$, represented in Figure 1.26 and for their division $z = y/x$ ($x \neq 0$), represented in Figure 1.27.

Also the elementary functions x^α, $\exp(x)$, $\log(x)$, and $\sqrt{x}$ of MATLAB are overloaded for treating fuzzy numbers as a list of intervals. Figures 1.28, 1.29, 1.30, and 1.31 list some examples.

1.2.3.7 The Malthus Law Revisited

As an application, we consider the same discrete Malthus law, which was considered for the interval case. We represent the initial value as a triangular fuzzy number

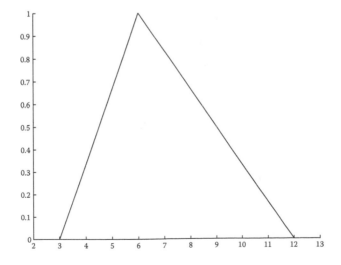

FIGURE 1.25 Product of the TFN (triangular fuzzy numbers) $[1, 2, 3]$ and $[2, 3, 4]$.

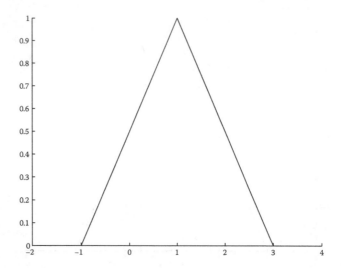

FIGURE 1.26 Difference of the TFN (triangular fuzzy numbers) $[2, 3, 4]$ and $[1, 2, 3]$.

X_0 comprised between 3 and 6 with peak value 4.5 and the growth constant (see Figure 1.32) λ as a triangular fuzzy number comprised between 0.95 and 1.05 with peak value 1.

With the MATLAB program fmalthus.m, we obtain, after 20 iterations, Figure 1.33.

1.2.4 Functions of Fuzzy Numbers

Let us consider a function $f : X \Rightarrow Y$ assumed to be bounded and continuous, with $X = X_1 \times \ldots \times X_n$.

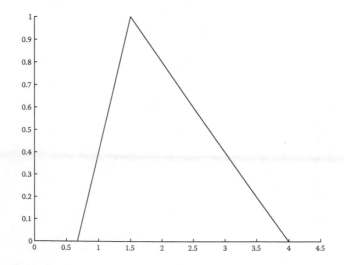

FIGURE 1.27 Division of the TFN (triangular fuzzy numbers) $[2, 3, 4]$ and $[1, 2, 3]$.

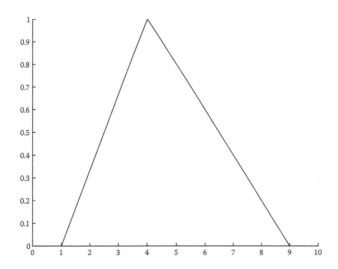

FIGURE 1.28 Square of the TFN $[1, 2, 3]$.

Let $A_1, A_2, \ldots, A_n$ be fuzzy sets belonging to $X_1, \ldots, X_n$, respectively. If the fuzzy sets $A_1, A_2, \ldots, A_n$ are normal and convex, and the function f has some appropriate properties, then also the set B, the range in Y of the function f, will be normal and convex, and therefore a fuzzy number.

For a general α-*level*, the problem is to compute the real interval:

$$B_\alpha = f(A_{1\alpha}, \ldots, A_{n\alpha}) = [y^L, y^R] \tag{1.86}$$

for each level α in $[0, 1]$, where y^L, y^R represent the global minimum and maximum of f in the space

$$X_\alpha = X_{1\alpha} \times \cdots \times X_{n\alpha} \tag{1.87}$$

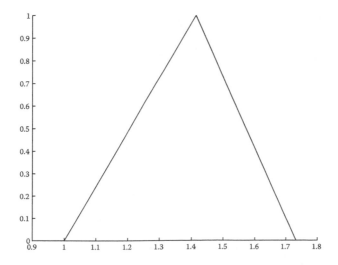

FIGURE 1.29 Square root of the TFN $[1, 2, 3]$.

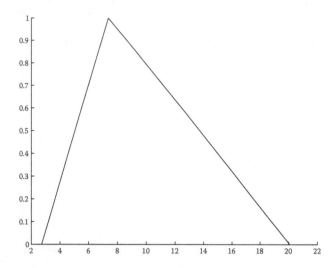

FIGURE 1.30 Exponential of the TFN $[1, 2, 3]$.

It is obvious that the number of α-*levels* used in the discretization of the interval $[0, 1]$ affects the accuracy of the approximation of the resulting fuzzy number.

From the above properties, it is apparent that the calculation of a fuzzy function can be reduced to the calculation of an interval function for the given number of α-levels. Therefore, the same problems of redundancy arise as with the interval arithmetic. These could be avoided if one utilizes the (computationally costly) global optimization algorithms. One of the most popular algorithms is the extension to the fuzzy case (i.e., for each α-level) of the *interval weighted average*, IWA, *algorithm*, which in this case is called *fuzzy weighted algorithm*, FWA in short. In order to be applicable, as in the interval case, it requires the function f to be monotonic in each argument.

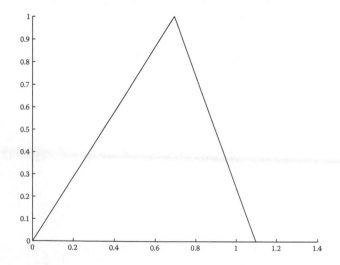

FIGURE 1.31 Logarithm of the TFN $[1, 2, 3]$.

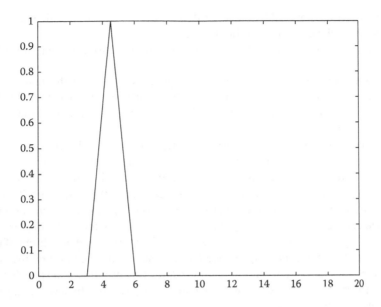

FIGURE 1.32 Initial value for X_0 in the discrete Malthus law.

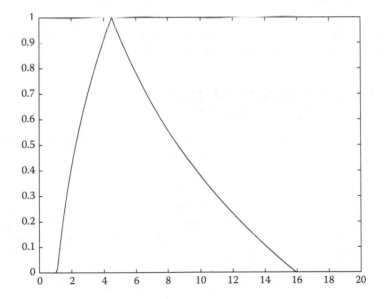

FIGURE 1.33 Malthus law with fuzzy numbers; $X_0 = TFN(3, 4.5, 6)$, $\lambda = TFN(0.95, 1, 1.05)$, after 20 iterations.

Other cases are covered by the algorithm of Yang et al. [101], which depends on the structure of the domain X of the function $f : X \Rightarrow Y$.

The domains are classified as uniform (or quasi-uniform) and nonuniform [101]:

X **is a uniform space** if X_α has no subspaces in which all the first partial derivatives of the function (assumed to be differentiable) vanish. In this case, one can apply the IWA algorithm to each α-level, that is, the FWA algorithm.

X **is quasi-uniform** if X_α has subspaces where the first partial derivatives of the function, $\partial_i f$, vanish but $\partial_i f$ do not depend on the variable x_i, $i = 1, \ldots, n$. Also in this case, the function f takes its extremal values at the corner points. Therefore, also in this case, one can apply the FWA algorithm as follows. One evaluates the n partial derivatives $\partial_i f$, and for each variable, one chooses x_i^L if $\partial_i f > 0$, otherwise x_i^R as a basis for evaluating y^L. In this way, one obtains y^L and y^R, respectively minimum and maximum of the function on the chosen x_i. Notice that it is sufficient to determine the sign of the derivatives only at the lowest α-level because then, the space being quasi-uniform, the sign of each derivative will remain the same at all the other α-levels.

X **is a nonuniform space** if the function can have extremal points within the domain. The algorithm then finds a list of points, called poles, which are candidates to be the extremal points of the function. In the end, it is equivalent to solving a nonlinear optimization problem.

All these algorithms can be easily parallelized on the basis of assigning the computation at each α-level to a different processor [19].

1.3   SEMANTICS OF INTERVAL, FUZZY, AND POSSIBILISTIC ENTITIES

There is often confusion about the differences between fuzzy and possibilistic entities. Fuzzy and possibilistic entities have a different development from first principles. Fuzzy set and possibility theory were defined and developed by L. Zadeh, beginning with [103] and subsequently [104], [105], [106], [107], and [108].

Associated with fuzzy sets, measures and membership functions are two other types of measures and ensuing distributions — possibility and necessity measures and distributions. Whereas fuzzy measures quantify the uncertainty of gradualness, possibility and necessity measures are ways to quantify uncertainty of lack of information. Books and articles are available that develop possibility theory (see, for example, [31], [27], [96], [62]). What is of interest here for fuzzy and possibilistic mathematical analysis that is developed, is what is called *quantitative* possibility theory [30]. In particular, possibility theory may be derived in any one of the following ways:

1. via *normalized* fuzzy sets (see [105], [106], and [107]),
2. axiomatically from *fuzzy measures* g that satisfy $g(A \cup B) = \max\{g(A), g(B)\}$ (see [32] and [63], for example),

3. via belief functions of Dempster-Shafer theory whose focal elements are normalized and nested (see [62]), or

4. by construction via nested sets with *normalization*, for example, nested *α-level* sets (see [58]).

Of special interest are the third and fourth approaches since they lead directly into a quantitative possibility theory, though the first points the way and lays the foundation. It will be assumed that the possibilistic/necessity measures and distributions that are used herein are constructed according to third or fourth approaches.

Fuzzy and possibilistic entities have different meanings, **semantics**. Fuzzy and possibility uncertainty model different entities and the associated solution methods are different as will be seen. Fuzzy entities, as is well known, are sets with nonsharp boundaries in which there is a transition between elements that belong and elements that do not belong to the set. Possibilistic entities are those that exist but the evidence associated with whether or not a particular element is that entity is incomplete, missing, or not available. Quantitative possibility distributions constructed from first principles require nested sets (see, for example, [58]) and normalization. Possibility distributions are normalized since their semantics are tied to existent entities. Normalization is not required of fuzzy membership functions. Thus, *not all fuzzy sets can give rise to possibility distributions.* That is, even though Zadeh's original development of possibility theory was derived from fuzzy sets, possibility theory is different from fuzzy set theory.

Possibilistic distributions (of fuzzy numbers, for example) encapsulate the best estimate of the possible values of an entity, given the available information. Fuzzy membership function values (of fuzzy numbers) describe the degree to which an entity is that value. Note that if the possibility distribution at x is 1, this signifies that the best evidence available indicates x is the entity that the distribution describes. On the other hand, if the fuzzy membership function value at x is 1, x is *certainly* the value of the entity that the fuzzy set describes. Thus, the nature of mathematical analysis in the presence of fuzzy and possibilistic uncertainties is quite different in semantics.

The most general form of possibility theory sets up an order among variables with respect to their being an entity. The *magnitudes* associated with this ordering have no significance other than an indication of order. Thus, if $possibility_A(x) = 0.75$ and $possibility_A(y) = 0.25$, all that can be said is that the evidence is stronger that x is the entity A than y. One *cannot* conclude that x is three times more likely to be A than y is. This means that for mathematical analysis, if the possibility distributions were constructed using the most general assumptions, comparisons among several distributions are restricted (to merely order). For the most general possibility theory, setting the possibility level to be greater than or equal to a certain fixed value α, $0 \leq \alpha \leq 1$, does not have the same meaning as setting a probability to be at least α. In the former case, the α has no inherent meaning (other than if one has a $\beta > \alpha$, one prefers the decision that generated β to that which generated α) whereas for the case of probability or quantitative possibility, the *value* of α is meaningful. The third and fourth derivations of possibility theory lead to quantitative possibility theory.

An alternative approach to possibility theory is as a system of upper and lower distributions bounding a given, yet unknown probability. That is, given a measurable set A, $Nec(A) \leq prob(A) \leq Pos(A)$ bounds the unknown probability of the event A so that $Pos(A) \leq \alpha$ guarantees that $prob(A) \leq \alpha$. If the possibilistic entities are constructed from this perspective, then their α-levels are numerically meaningful beyond simply being an ordering. This is the method developed in [58].

2 Interpolation with Data Containing Interval, Fuzzy, and Possibilistic Uncertainty

Marcello Anile and Salvatore Spinella

CONTENTS

2.1 GENERAL DEFINITIONS

2.1.1 MODELING OBSERVATIONS

Let O be a sequence of n observational data in a domain $X \subseteq R^2$ in the form

$$O = \{(x_1, y_1, Z_1), \ldots, (x_i, y_i, Z_i), \ldots, (x_n, y_n, Z_n)\} \qquad (2.1)$$

with

$$(x_i, y_i) \in X \quad Z_i = \{z_{i,1}, \ldots, z_{i,m_i}\} \qquad (2.2)$$

where $z_{i,j} \in R$ represents the jth observation at the point (x_i, y_i).

As previously discussed in the introduction, a first approach to data reduction is to model the observation set Z_i with the interval I_i defined by the minimum and maximum values of the observations at the point (x_i, y_i) [40]:

$$I_O = \{(x_1, y_1, I_1), \ldots, (x_i, y_i, I_i), \ldots, (x_n, y_n, I_n)\} \qquad (2.3)$$

where I_i is the interval

$$I_i^l = \min_{1 \leq j \leq m_i} z_{i,j} \quad I_i^u = \max_{1 \leq j \leq m_i} z_{i,j}$$

For some applications, this reduction is *too coarse because it does not take into account the distribution and the quality of the data* Z_i at (x_i, y_i). If the data set Z_i *comprises a large number of observations* $z_{i,j}$, a representation of Z_i in terms of a *probability density* can be achieved and then a stochastic model can be fitted to the data in the form of a stochastic surface. However, in general, this approach requires special assumptions on the *joint probability density* of the data distribution, which are difficult to check from the data themselves. Furthermore, the interrogation of the stochastic surface model can be done, in general, only through Monte Carlo simulation, which is both noisy and computationally very expensive. Finally, this approach cannot be pursued when the data set contains only few data and the data quality is judged subjectively from the observer.

For all these reasons, another approach has been introduced [3, 72] that represents the datum Z_i with an appropriately constructed fuzzy number that reflects *both the data distribution and their quality*. Here, fuzzy numbers [60] are defined as maps that associate to each *presumption level* $\alpha \in [0, 1]$ a real interval A_α such that

$$\alpha'' > \alpha' \Rightarrow A_{\alpha''} \subseteq A_{\alpha'} \tag{2.4}$$

and the latter property is formally called *convex hull property*. A regularity property such as upper semicontinuity is usually also invoked. The core of a fuzzy number is the set of values with membership equal to 1. When the core has only one element, the latter is called modal value. The support of the fuzzy number is an interval where the membership function is positive. In this representation, fuzzy numbers are viewed as the natural generalization of intervals and *intervals can be viewed as fuzzy numbers whose core equals its support*.

By utilizing one of several methods for constructing fuzzy set membership functions ([61], [48]) from Z_i, one can represent the n observational data as

$$F_O = \{(x_1, y_1, \tilde{z}_1), \ldots, (x_i, y_i, \tilde{z}_i), \ldots, (x_n, y_n, \tilde{z}_n)\} \tag{2.5}$$

where $\tilde{z}_i \in F(R)$ is the fuzzy number representing the observations at the point (x_i, y_i).

Arithmetic operations among reals are extended to operations among fuzzy numbers by utilizing first the *extension principle* of interval arithmetic [54].

DEFINITION 2.1
(Operation $\otimes$ between intervals). Given two intervals $[a, b]$ and $[c, d]$, the operation $[a, b] \otimes [c, d]$ is defined formally by

$$[a, b] \otimes [c, d] = \min_{[e,f]}(\{x \otimes y | x \in [a, b], y \in [c, d]\} \subseteq [e, f]) \tag{2.6}$$

Likewise, the operations among fuzzy numbers are introduced by the following definition ([60], [2]).

DEFINITION 2.2
(Operation $\otimes$ between fuzzy numbers). The operation $F_1 \otimes F_2$ associates with the fuzzy numbers $F_1 = \{A_\alpha\}_{\alpha \in [0,1]}$ and $F_2 = \{B_\alpha\}_{\alpha \in [0,1]}$; the fuzzy number $F = \{C_\alpha\}_{\alpha \in [0,1]}$ such that $\forall \alpha \in [0, 1]$ one has $C_\alpha = A_\alpha \otimes B_\alpha$

Fuzzy arithmetic, which is based on interval arithmetic, is *conservative* with respect to the uncertainty of the data in the sense that, by utilizing the min-max operations, it yields under/overestimation for each α-cut.

2.1.2   REAL B-SPLINES

For an introduction to the theory of B-splines, see the book by Lancaster and Saskaulas [63]. A B-spline function of order h is a piecewise polynomial function $f(t)$: $[t_0, t_m] \rightarrow R$ of degree at most $h - 1$. The sequence of knots of a B-spline is a nondecreasing sequence of real numbers: $(t_0, \ldots, t_i, \ldots, t_m)$. Let $k = m - (2h - 1)$ where h is the order of the B-spline. The knots of the subsequence $(t_{h-1}, \ldots, t_{k+h-1})$ are the inner knots of the B-spline. $f(t)$ is C^∞ in $[t_0, t_m]$, but at a knot of multiplicity p, $f(t)$ is only C^{h-p-1}. A B-spline of order h on a sequence of $m = k + 2h - 1$ knots is a linear combination

$$f(t) = \sum_{i=0}^{k+h-1} c_i B_{i,h}(t) \tag{2.7}$$

where c_i are the control coefficients, or control points, and $B_{i,h}(t)$ are the basis functions of the B-splines of order h defined recursively by [24]

$$B_{i,1}(t) = \begin{cases} 1 & \text{if } t_i \leq t \leq t_{i+1} \\ 0 & \text{otherwise} \end{cases}$$

$$B_{i,h}(t) = \frac{t - t_i}{t_{i+h-1} - t_i} B_{i,h-1}(t) + \frac{t_{i+h} - t}{t_{i+h} - t_{i+1}} B_{i+1,h-1}(t) \quad \text{for } h > 1 \tag{2.8}$$

A real B-spline surface is a tensor product of B-splines defined on a rectangular domain of $M \times N$ control points $\{c_{i,j}\}_{i=0,\ldots,M-1, j=0,\ldots,N-1}$:

$$f(u, v) = \sum_{i=0}^{M-1} \sum_{j=0}^{N-1} c_{i,j} B_{i,h}(u) B_{j,h}(v) \tag{2.9}$$

The basis functions of real B-splines are all positive and guarantee, among other things, the *convex hull* property, that is, that $f(t)$ is contained, for all t in the convex envelope of the control points.

2.1.3   INTERVAL B-SPLINES

The concept of real B-spline can be easily extended to the space $I(R)$ of real intervals [40].

DEFINITION 2.3
(Interval B-spline). An interval B-spline $F(t)$ relative to the knot sequence $(t_0, t_1, \ldots, t_m)$, $m = k + 2(h - 1)$, is a function of the kind $F(t) : R \rightarrow I(R)$ defined as

$$F(t) = \sum_{i=0}^{h+h-1} I_i B_{i,h}(t) \tag{2.10}$$

where the control coefficients I_i are real intervals and $B_{i,h}(t)$ real B-spline basis functions.

The generalization in 2D of an interval B-spline relative to a rectangular grid of $M \times N$ knots is

$$f(u, v) = \sum_{i=0}^{M-1} \sum_{j=0}^{N-1} I_{i,j} B_{i,h}(u) B_{j,h}(v) \tag{2.11}$$

The construction of an interval B-spline will be considered under the more general case of a fuzzy B-spline, since intervals can be considered as a special case of fuzzy numbers.

2.1.4   FUZZY B-SPLINES

Similar to what is done for intervals in [40], we introduce fuzzy B-spline as follows [3]:

DEFINITION 2.4
(B-spline fuzzy) A fuzzy B-spline $F(t)$ relative to the knot sequence $(t_0, t_1, \ldots, t_m)$, $m = k + 2(h - 1)$ is a function of the kind $F(t) : R \to F(R)$ defined as

$$F(t) = \sum_{i=0}^{h+h-1} F_i B_{i,h}(t) \tag{2.12}$$

where the control coefficients F_i are fuzzy numbers and $B_{i,h}(t)$ real B-spline basis functions.

Notice that Definition 2.4 is consistent with the previous definitions and more precisely for any t, $F(t)$ is a fuzzy number, that is, it verifies the convex hull property given by Equation (2.4):

$$\alpha'' > \alpha' \Rightarrow F(t)_{\alpha''} \subseteq F(t)_{\alpha'}$$

because the B-spline basis functions are nonnegative.

The generalization in 2D of a fuzzy B-spline relative to a rectangular grid of $M \times N$ knots is

$$f(u, v) = \sum_{i=0}^{M-1} \sum_{j=0}^{N-1} F_{i,j} B_{i,h}(u) B_{j,h}(v) \tag{2.13}$$

with the same properties as described above. Similar considerations in the more general framework of fuzzy interpolation can be found in [71].

2.2   CONSTRUCTING FUZZY B-SPLINE SURFACES

Let us consider a sequence of fuzzy numbers representing the observations in Equation (2.5). If a fuzzy B-spline $F(u, v)$ on a rectangular grid $G \supseteq X$ of $M \times N$ knots approximates F_O given by Equation (2.5) then

$$\forall \alpha \in [0, 1] \quad [\tilde{z}_i]_\alpha \subseteq [F(x_i, y_i)]_\alpha \quad i = 1, \ldots, n \qquad (2.14)$$

and furthermore, one must also have

$$\int_G \sum_{i=0}^{M-1} \sum_{j=0}^{N-1} \left([F_{i,j}^u]_\alpha - [F_{i,j}^l]_\alpha \right) B_{i,h}(u) B_{j,h}(v) \, du \, dv$$

$$\leq \int_G \sum_{i=0}^{M-1} \sum_{j=0}^{N-1} \left([Y_{i,j}^u]_\alpha - [Y_{i,j}^l]_\alpha \right) B_{i,h}(u) B_{j,h}(v) \, du \, dv$$

$$\forall \{Y_{i,j}\}_{i=0,\ldots,N-1, j=0,\ldots,M-1} \in F(R) \quad \forall \alpha \in [0, 1] \qquad (2.15)$$

where $[F^l]_\alpha$ and $[F^u]_\alpha$ indicate, respectively, the lower and upper bounds of the interval representing the fuzzy number α-level. More precisely, for each presumption α-level, the volume encompassed by the upper and lower surfaces of the fuzzy B-spline is the smallest. These definitions are the generalization to the fuzzy case of the corresponding interval ones [40].

Notice that the integral in a rectangular domain of a real B-spline is the following linear expression:

$$\int_D \sum_{i=0}^{M-1} \sum_{j=0}^{N-1} P_{i,j} B_{i,h}(u) B_{j,h}(v) \, du \, dv = \sum_{i=0}^{M-1} \sum_{j=0}^{N-1} P_{i,j} \frac{(t_{i+h} - t_i)(s_{j+h} - s_j)}{h^2} \qquad (2.16)$$

where obviously $\{(t_i, s_j)\}_{i=0,\ldots,M+h-1, j=0,\ldots,N+h-1}$ are the grid knots.

Therefore, given the set F_O of observations in Equation (2.5) and a finite number $P+1$ of presumption levels $\alpha_0 > \alpha_1 > \cdots > \alpha_P$, the construction of a fuzzy B-spline requires the solution of the following *constrained optimization problem*:

$$\begin{cases} \min \sum_{k=0}^{P} \sum_{i=0}^{M-1} \sum_{j=0}^{N-1} \left([F_{i,j}^u]_{\alpha_k} - [F_{i,j}^l]_{\alpha_k} \right) \frac{(t_{i+h}-t_i)(s_{j+h}-s_j)}{h^2} \\ [F_{i,j}^l]_{\alpha_0} \leq [F_{i,j}^l]_{\alpha_1} \leq \cdots \leq [F_{i,j}^l]_{\alpha_P} \leq [F_{i,j}^u]_{\alpha_P} \leq \cdots \\ \cdots \leq [F_{i,j}^u]_{\alpha_1} \leq [F_{i,j}^u]_{\alpha_0} \quad i = 0 \ldots M-1 \quad j = 0 \ldots N-1 \\ \sum_{i=0}^{M-1} \sum_{j=0}^{N-1} [F_{i,j}^u]_{\alpha_k} B_{i,h}(x_r) B_{j,h}(y_r) \geq [z_r^u]_{\alpha_k} \quad r = 1, \ldots, n \quad k = 0, \ldots, P \\ \sum_{i=0}^{M-1} \sum_{j=0}^{N-1} [F_{i,j}^l]_{\alpha_k} B_{i,h}(x_r) B_{j,h}(y_r) \leq [z_r^l]_{\alpha_k} \quad r = 1, \ldots, n \quad k = 0, \ldots, P \end{cases}$$

$$(2.17)$$

With the variable transformation

$$F_{i,j}^l = \overline{F}_{i,j}^l - \overline{\overline{F}}_{i,j}^l \quad F_{i,j}^u = \overline{F}_{i,j}^u - \overline{\overline{F}}_{i,j}^u$$

and the further constraints

$$\overline{F}_{i,j}^l \geq 0 \quad \overline{\overline{F}}_{i,j}^l \geq 0 \quad \overline{F}_{i,j}^u \geq 0 \quad \overline{\overline{F}}_{i,j}^u \geq 0$$

such a problem can be treated as a *linear programming problem*.

2.2.1 REMARKS

For the problem in Equation (2.17), one notices that

1. The objective function minimizes the uncertainty of the representation.
2. The first group of constraints represents the convex hull property.
3. The second group of constraints is a consistency requirement for the upper bounds of the intervals to be greater than the upper bounds of the data at the data points.
4. The third group of constraints is a consistency requirement for the lower bounds of the intervals to be lower than the lower bounds of the data at the data points.

By considering a grid of size $M \times N$, T fuzzy observations, and P presumption levels, the dimensions of the problem are of $4MNP$ variables and $2(P-1)MN + 2T$ constraints [70].

2.3 FUZZY KRIGING AND BOUNDARY CONDITION

In the previous paragraph, we expounded the method for constructing a fuzzy surface approximating, in a well defined way, the fuzzy numbers representing the data sets. The quality of this approximation will deteriorate the farther one is from the sites of the data sets. Therefore, one expects the constructed approximation not to be very satisfactory at the border of the domain within which the data sites are comprised. To remedy such drawbacks, one has to introduce further information regarding the decay of the quantity of interest away from the domain.

A simple approach would be to assume that the quantity of interest decays to zero outside the boundary but this is hardly justifiable. However, a better treatment would be to construct fictitious observation data just outside the boundary by utilizing statistical kriging. The latter approach is more realistic because, in some sense, it amounts to an extrapolation driven by the data.

By ordinary spatially distributed data and with stationary hypothesis of the distribution, the *kriging* combines the available data by weights in order to construct an *unbiased* estimator with minimum variance. Likewise, such approach is extended to the fuzzy case [25]. Given N spatially distributed fuzzy data $\{(x_1, y_1, \widetilde{z}_1), \ldots, (x_i, y_i, \widetilde{z}_i), \ldots, (x_n, y_n, \widetilde{z}_n)\}$, one looks for an estimator constructed by a linear combination of weights λ_i to evaluate the distribution at the point (x, y) like

$$\widetilde{Z}^* = \sum_{i=1}^{N} \lambda_i \widetilde{z}_i \qquad (2.18)$$

Notice that the above interprets fuzzy data $\widetilde{z}_i$ and estimator $\widetilde{Z}^*$ like *random fuzzy number*. In order to construct this estimator, the following hypothesis must be satisfied

(K1) $E(\widetilde{Z}^*) = E(\widetilde{Z}) = E(\widetilde{Z}(x+h, y+k)) = r$

(K2) $Ed(\widetilde{Z}^*, \widetilde{Z})^2$ must be minimized $\qquad (2.19)$

(K3) $\lambda_i \geq 0, \quad i = 1, \ldots, N$

The first condition (K1) implies $\sum_{i=1}^{N} \lambda_i = 1$; instead, the last one (K3) guarantees the estimator stands in the cone of random fuzzy number generated by the data.

It can be proved that

$$Ed(\tilde{Z}^*, \tilde{Z})^2 = \sum_{i,j=1}^{N} \lambda_i \lambda_j \sum_\alpha C_\alpha(x_i, y_i, x_j, y_j)$$
$$-2 \sum_{i=1}^{N} \lambda_i \sum_\alpha C_\alpha(x_i, y_i, x, y) \qquad (2.20)$$
$$+ \sum_\alpha C_\alpha(x, y, x, y)$$

where C_α is a positive defined function that represents a *covariance*.

The minimization of such function together to the hypotheses (K1) and (K3) leads to a constrained minimization problem. It is solvable formulating everything in terms of Kuhn-Tucker conditions by the following theorem.

THEOREM 2.1
Let $\tilde{Z}^*$ be an estimator for $\tilde{Z}$ of the form $\tilde{Z}^* = \sum_{i=1}^{N} \lambda_i \tilde{z}_i$. Suppose the conditions (K1) and (K3) are satisfied and the matrix defined by $\Gamma_{ij} = \sum_\alpha C_\alpha(x_i, y_i, x_j, y_j)$, $i, j = 1, \ldots, N$ is strictly positive defined. Then there exists a unique linear unbiased estimator $\tilde{Z}^*$ satisfying the (K2) condition. Moreover, the weight satisfies the following system

$$\sum_{i=1}^{N} \Gamma_{ij} \lambda_{ij} - L_j - \mu = \sum_\alpha C_\alpha(x_i, y_i, x, y)$$
$$\sum_{i=1}^{N} \lambda_i = 1$$
$$\sum_{i=1}^{N} L_i \lambda_i = 0 \qquad (2.21)$$
$$L_i, \lambda_i \geq 0, \ i = 1, \ldots, N$$

The residual is

$$\sigma^2 = \sum_\alpha C_\alpha(x, y, x, y) + \mu - \sum_{i=1}^{N} \lambda_i \sum_\alpha C_\alpha(x_i, y_i, x, y) \qquad (2.22)$$

The above problem defines the weights λ_i and it can be solved using a method to manipulate the constraints in Equation (2.21) like the active set method [42].

3 Introduction to Geographical Information Systems

Cidália Fonte, Jorge Santos, and Gil Gonçalves

CONTENTS

3.1 GEOGRAPHICAL INFORMATION SYSTEMS

3.1.1 INTRODUCTION

Geographical information system (GIS) is an information system where the geographical locations of the phenomena are stored, enabling the retrieval and processing of geographical information.

3.1.1.1 Data Models

Traditionally, the geographical information is modeled considering the *object data model* or the *field data model*. The phenomena that exist only on a limited region of the geographical space are represented by objects whose position in the geographical space represents their location. Depending on the type of phenomena to be represented, they may be modeled as points, lines, or areas (see Figure 3.1). This data model is known as the object data model and is appropriate to represent, for example, roads, administrative regions, or crops.

When the phenomenon to be represented exists in all geographical space, and what is to be represented is its variation over the geographical space, the phenomenon is modeled using a surface $z = f(x, y)$, where the coordinates x and y represent the

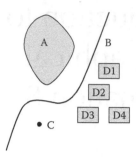

FIGURE 3.1 Representation of phenomena using the object data model. Objects A, D1, D2, D3, and D4 represent areal entities; object B represents a linear entity and object C a point entity.

geographical planimetric position and z the value of the phenomenon at that location (see Figure 3.2). This data model is known as the field data model and is appropriate to represent phenomena such as altitude, atmospheric pressure, or rainfall.

In addition to these traditional data models, a hybrid model may be considered, based on the use of fuzzy geographical entities (FGEs). This approach enables the representation of objects with surfaces, gathering therefore the important characteristics of both traditional data models. It may be used, for example, to represent GEs with ill-defined boundaries or whose position varies continuously in the geographical space. Further details about the construction and processing of FGEs are given in chapter 4.

3.1.1.2   Data Structures

For the digital representation of the phenomena, two data structures are traditionally considered: the *vector data structure* and the *raster data structure*. The vector data

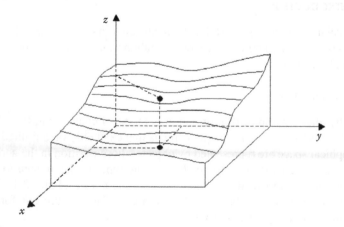

FIGURE 3.2 Representation of a phenomenon using the surface data model. The coordinates x and y represent the geographical position and coordinate z the value and concentration of amplitude of the phenomenon.

structure uses primitives as points, lines, and areas, and the geographical space is considered continuous. A point is represented by its coordinates (x, y); a line is represented by a sequence of points, which are usually connected by straight segments, or, less frequently, by curves; and an area is represented by a closed line corresponding to its boundary. The vector data structure is usually associated to the representation of objects, but it can also be used to represent surfaces, using, for example, contours, profiles, or a tessellation of regions such as the Delaunay triangles or the Voronoi polygons (see Figure 3.3).

The raster data structure is characterized by a grid of cells, usually square, known as pixels. Since only a value of each phenomenon is associated to each cell, in the raster

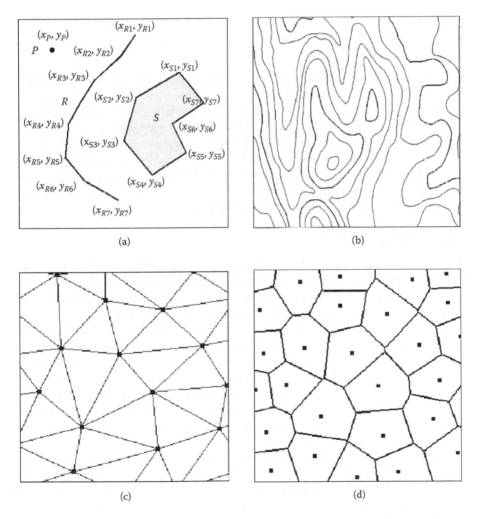

(a)   (b)

(c)   (d)

FIGURE 3.3 (a) Representation of entities using the vector data structure. Representation of surfaces using the vector data structure: with (b) contours; (c) Delaunay triangles; (d) Voronoi polygons.

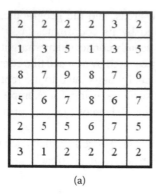
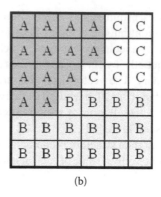

2	2	2	2	3	2
1	3	5	1	3	5
8	7	9	8	7	6
5	6	7	8	6	7
2	5	5	6	7	5
3	1	2	2	2	2

(a)

A	A	A	A	C	C
A	A	A	A	C	C
A	A	A	C	C	C
A	A	B	B	B	B
B	B	B	B	B	B
B	B	B	B	B	B

(b)

FIGURE 3.4 (a) A surface represented with the raster data structure. (b) Entities (objects) represented with the raster data structure.

data structure, the geographical space is considered discrete. This type of structure adapts well to the representation of surfaces, even though its resolution is conditioned to the dimension of the cells. The raster data structure may also be used to represent objects. Each object is represented explicitly by a set of contiguous cells, corresponding to the same attribute value, such as soil type or slope classes (see Figure 3.4).

A tessellation is a representation where the geographical space is partitioned into a finite number of contiguous elementary regions, which may be regular or irregular. To each elementary region, a value of the phenomenon to be represented is associated. The raster data structure is a particular case of a tessellation.

3.1.1.3 Analysis in GIS

An important characteristic of GISs is their capability of incorporating functions to perform operations with the stored geographical information, providing analysis capabilities. A great variety of operations may be executed, which can be grouped into two types: *operations with attributes* and *spatial operations* [18].

The operations with attributes involve only the nonspatial attributes of the stored geographical information. Examples of these operations are querying the database about the owner or soil type of a parcel identified by the parcel number. These queries do not involve information about the spatial location of the parcel. The spatial operations use the data about the geographical location of the information, and the analysis performed with these operations is called *spatial analysis*. Among this type of operation are, for example, the computation of areas or the identification of the parcels located within 500 meters of a road.

Spatial analysis often requires the consecutive use of several operators, which may process the information initially stored in the GIS or information generated through the previous application of other operators. Some operators just retrieve information from the GIS, such as the identification of the parcels with certain characteristics, although many operators generate new attributes and new geographical information ([6], [18], [93]).

3.1.2 THE OBJECT DATA MODEL — GEOGRAPHICAL ENTITIES

In object-based GISs, the geographical information is represented by GEs. These entities are characterized by an attribute and a spatial location. The attribute specifies the phenomenon represented by the entity, such as building, forest, parcel, or more complex attributes, such as rural region with a slope between 10 and 20 percent, good solar exposure, and high levels of humidity. The spatial location of a geographical entity is specified by a point, a line, or a region and may be represented either in the vector or in the raster data structure.

A group of GEs characterized by the same attribute is a *class of GEs*, or simply a *class*.

Construction of GE

Geographical entities may be built using several methods and with several types of data. The sources of geographical data may be grouped into primary and secondary [57]. The primary sources correspond to the cases where the data are collected directly from the geographical space, such as with topographic or photogrammetric surveys. The secondary methods correspond to the use of information already collected and processed, eventually for other aims, which may be in digital or analog format.

Basically two approaches may be considered to construct GEs:

Attribute-location approach
This approach consists in identifying the attribute that characterizes the geographical entity, and determining its geographical location, that is, identifying the regions of the geographical space where that attribute can be found. One way to determine the geographical location is to identify the attribute in the terrain, aerial photos, remote sensing images, or even paper maps, and determine the position of the point, line, area, pixels, or cells corresponding to that attribute. The information may be collected in the vector data structure, as in the case of topographic surveys, vectorization of paper maps, or aerial photos, or in the raster data structure, for example, when paper maps are digitized to the raster data structure using a scanner. However, in most cases, the construction of GE in the raster data structure results from processing of information already in the raster data structure, such as the classification of remote sensing images.

This last case may be included in a more vast procedure, named here *classification of tessellations*, which consists in classifying tessellations representing one or several phenomena or attributes, named *base attributes*, generating new attributes, named *derived attributes*. The tessellations are formed by elementary regions r_i, which may be cells in a raster data structure or any other regions, such as, for example, irregular polygons corresponding to classes of slope. Each derived attribute corresponds to a set of values of the base attribute and characterizes a class of GEs. For example, the attribute "forest" may correspond to a set of levels of radiance in a multispectral image obtained by sensors installed in artificial satellites, or the attribute "mountain"

Values of the base attributes

Derived $\overbrace{\{1, 2, 3, 4, 5, 6, 7, 8, 9, 10\}}$
attributes   A   B   C

FIGURE 3.5 Correspondence between the values of the base attributes and the values of the derived attributes.

to a set of values of terrain altitude. In these cases, the GEs are obtained aggregating contiguous elementary regions to which the same derived attribute was assigned during the classification. For example, in Figure 3.6, the derived attributes were defined as a function of the base attributes, where derived attribute A corresponds to the interval [1, 3] of the base attributes, attribute B to the interval [4, 7], and attribute C to the interval [8, 10] (see Figure 3.5).

Location-attribute approach

The reverse of the previous approach may also be used to build GE. In this case homogeneous regions are identified in the geographical space. When their position is determined, the attribute characterizing them will be identified. The location-attribute approach may also be used either with the vector data structure or the raster data structure. In the former, regions are identified, for example, in aerial photos or high resolution remote sensing images, and the subsequent identification of the corresponding phenomena is done. In the raster data structure, the classification of remote sensing images with the unsupervised classification method is an example of this approach.

3.1.3 THE FIELD DATA MODEL

3.1.3.1 The Field Data Model

The field-based approach conceptualizes the geographic space as a collection of fields. Each field defines the spatial variation of an attribute of interest as a mathematical

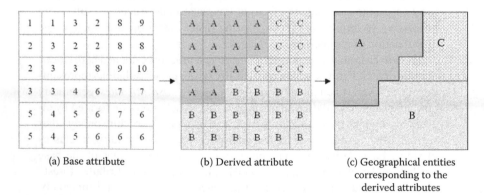

(a) Base attribute   (b) Derived attribute   (c) Geographical entities corresponding to the derived attributes

FIGURE 3.6 Geographical entities obtained aggregating contiguous cells to which the same derived attribute was assigned during the classification of the base attribute.

function $Z = f(x, y)$. This function is defined on a given region of the geographic space and assigns to every element of that region a corresponding field attribute value from the attribute domain. The field model usually assumes a partition of the geographic space into a finite tessellation of spatial entities. The tessellation can be regular, such as a grid of squares, or irregular, such as a network of triangles or Voronoi polygons. The values of the attribute domain are commonly classified using four scales of measurement: nominal, ordinal, interval, and ratio. Categorical or qualitative attributes are often expressed in nominal and ordinal scales. Quantitative attributes are reported in interval or ratio scales. The fields may be continuous, differentiable, discrete, and isotropic or anisotropic, with positive or negative autocorrelation. The basic mathematical model $Z = f(x, y)$ can be generalized to allow tridimensional spatial coordinates $\mathbf{x} = (x, y, z)$, time parameter t, and multivalued fields $\mathbf{Z} = (Z_1, Z_2, ..., Z_n)$, taking the form $\mathbf{Z} = f(\mathbf{x}, t)$.

3.1.3.2 Map Algebra Operations on Fields

Map algebra describes a set of field operations that are typically classified into local, focal, and zonal types [93]. These operations take as input one or more fields and generate a separate field as an output data layer. In a local operation, the value of the new field at any location $\mathbf{x}$ is dependent only on the values of the input field functions at that location. In focal operations, the value of the new field at a location $\mathbf{x}$ depends not only on the attribute of the input field but also on the attributes of these functions in a neighborhood of $\mathbf{x}$. Zonal operations take as input a value field and a zone field, and aggregate the values of the field over each zone.

4 Geographical Entities as Surfaces

Cidália Fonte

CONTENTS

4.1 INTRODUCTION

The construction of geographical entities (GEs) may be influenced by several types of errors and uncertainty that generate errors and uncertainty in the attributes and/or the geographical location of the entities. In other situations, the need to choose between the object data model and the field data model is a source of error, since some phenomena have characteristics of both data models and therefore are not adequately represented by either of them.

Fuzzy set theory incorporates some concepts that can be used to overcome some of the problems described above, modeling some types of uncertainty associated to geographical information as well as its heterogeneity. They enable the development of an alternative data model that integrates characteristics of the object and the field data models, where the geographical information is represented with fuzzy geographical entities (FGEs), which are geographical entities (objects) represented with surfaces.

An overview of the several sources of uncertainties and/or errors associated with GEs is presented in this chapter as well as methods to model their uncertainty or heterogeneity using fuzzy sets.

FGEs are an efficient way to represent the positional uncertainty of GEs, or a gradual variation between them, but their inclusion in geographical information systems (GISs) requires not only their construction but also the development of operators capable of processing them. Some basic operators that process FGEs have already been developed and are briefly reviewed in this chapter.

4.2   FUZZY GEOGRAPHICAL ENTITIES

Since this chapter addresses the representation of GEs as surfaces that represent the uncertainty or heterogeneity of the entities using fuzzy sets, through the construction of FGEs, only the types of errors and uncertainty that can be modeled with this approach are considered. A GE is characterized by an attribute and a geographical location. These two components of the GEs are intimately related, and therefore the errors and uncertainty associated to each of them are also related.

A FGE E_A, characterized by attribute "A," is a GE whose position in the geographical space is defined by the fuzzy set

$$E_A = \{(x, y) : (x, y) \text{ belongs to the GE characterized by attribute "}A\text{"}\}$$

with membership function $\mu_{E_A}(x, y) \in [0, 1]$ defined for every location in the space of interest. The membership value 1 represents full membership. The membership value 0 represents no membership, and the values in between correspond to membership grades to E_A, decreasing from 1 to 0.

The construction of FGEs is based on the construction of the membership function $\mu_{E_A}(x, y)$, so its construction is of prime importance, but it is also one of the main difficulties of using fuzzy sets ([37], [95], [110]). Since the grades of membership to a fuzzy set may have several semantic interpretations, they may be used in several contexts with different meanings and therefore their semantic interpretation should be identified.

Some methods to build FGEs are presented in the next section, and a semantic interpretation of the grades of membership is given for each case.

4.2.1 CONSTRUCTION OF FUZZY GEOGRAPHICAL ENTITIES

In chapter 3, section 3.1.2, two main approaches were presented to build GEs, the attribute-location approach and the location-attribute approach. In both approaches, several sources of uncertainty may be considered.

4.2.1.1 Attribute Definition

In the attribute-location approach, the first step is the definition of the attribute that characterizes the GE. This definition may just consist in the choice of a concept easily identified by humans on the ground, aerial photographs, or satellite images, such as buildings, roads, or rivers, or may depend upon a group of measurable characteristics (taking values on a scale Z), such as, for example, radiance levels, altitude, or slope. In this case, the attribute definition involves a series of choices to identify the appropriate characteristics, which, in some cases, may not be easily determined. For example, suppose the regions suitable to construct an infrastructure are the GEs characterized by slopes between 10 and 20 percent and good solar exposure. Why choose slopes between 10 and 20 percent and not 9 and 21 percent? What is a good solar exposure? Since these choices are sometimes subjective and/or difficult to define, it may be considered that, in some situations, there is uncertainty associated to the definition of the attributes that characterize the GEs.

When the attribute definition is just based on the choice of a concept, it is assumed that humans are able to identify it without any measured values. The attribute definition may be more or less precise, giving some instructions regarding, for example, what types of constructions are considered as buildings. With this approach, most difficulties will be encountered in the second phase of the construction of the GEs, that is, in the identification of the locations on the ground that correspond to the attribute, since the process is subjective. The location of the GE will depend upon the identification of the regions where the attribute occurs, and therefore any misinterpretation of its meaning will generate uncertainty in the position of the GE. So, the attributes should be defined with some detail to reduce any doubts that might occur during its identification on the terrain.

Where the attribute is defined by a group of values within one or several characteristics chosen to identify the attribute, the values z of scale Z that correspond to the attribute have to be chosen for each characteristic considered in the definition of the attribute. In most situations, these values correspond to an interval $Z_A = [\underline{z}, \overline{z}]$ (see Figure 4.1), and therefore, $\underline{z}$ and $\overline{z}$ have to be identified for each attribute. One source of uncertainty in this case is that the transition between the set of values that correspond to the attribute and the ones that do not may not be clear, and therefore the attribute may be more accurately defined if, instead of choosing a crisp set Z_A, a gradual transition between membership and nonmembership is allowed, using a fuzzy set $\widetilde{Z_A}$ (see Figure 4.2).

The construction of fuzzy set $\widetilde{Z_A}$ requires the assignment to every value z of a grade of membership to the set, that is, the definition of a membership function of the z values to the set.

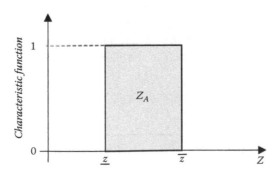

FIGURE 4.1 Set of Z values that define attribute A.

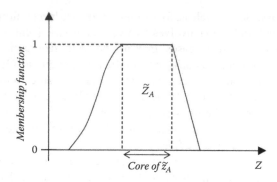

FIGURE 4.2 Fuzzy set $\widetilde{Z}_A$.

The semantic import approach used frequently to build membership functions (e.g., [14], [16], [74], [75], and [95]) considers this approach. Within this method, the grades of membership represent degrees of similarity.

A method to build the membership functions based on this semantic interpretation is presented in [46]. The similarity view interprets grades of membership as a quantification of the similarity between the observed characteristics and the ideal ones [92]. The first step to build the membership function is to identity the values that ideally correspond to the attribute, that is, to identity the core $Z_{A_{id}} = [\underline{z_{id}}, \overline{z_{id}}]$ of the fuzzy set. The second step corresponds to the computation of the degrees of similarity between the other z values and the ideal ones. These may be quantified considering the distance between those values and the extreme points of interval $Z_{A_{id}}$. The distance between value z and the core of fuzzy set $\widetilde{Z}_A$ (see Figure 4.3) is given by

$$d(z, Z_{A_{id}}) = \begin{cases} \min[d(z, \underline{z_{id}}), d(z, \overline{z_{id}})] & \Longleftarrow z \notin [\underline{z_{id}}, \overline{z_{id}}] \\ 0 & z \in [\underline{z_{id}}, \overline{z_{id}}] \end{cases} \tag{4.1}$$

where $d(z, \underline{z_{id}})$ and $d(z, \overline{z_{id}})$ are, respectively, the Euclidean distance between z and the extreme points of interval $Z_{A_{id}}$ (see Figure 4.4).

The degrees of membership of the z values to $\widetilde{Z}_A$ are computed considering a function f such that $\mu_{\widetilde{Z}_A}(z) = f[d(z, Z_{A_{id}})]$. Function f is application dependent and translates the degree of variation of the membership grades with the distance.

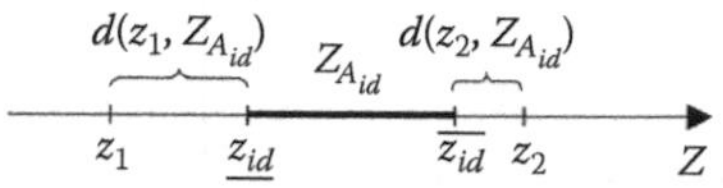

FIGURE 4.3 Distances $d(z_1, Z_{A_{id}})$ and $d(z_2, Z_{A_{id}})$ are, respectively, the distances of values z_1 and z_2 to the set of ideal values corresponding to attribute A, that is, to $Z_{A_{id}} = [\underline{z_{id}}, \overline{z_{id}}]$.

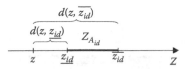

FIGURE 4.4 Distances $d(z, \underline{z_{id}})$ and $d(z, \overline{z_{id}})$ are, respectively, the Euclidean distance of z to the extreme points of interval $Z_{A_{id}}$.

It must be a decreasing function such that

$$
\begin{cases}
f(0) = 1 \\
\lim_{d(z, Z_{A_{id}}) \to +\infty} f[d(z, Z_{A_{id}})] = 0
\end{cases}
\tag{4.2}
$$

In most applications, positive memberships are considered only for distances smaller than a predefined value $d_{\max}$. This generates a fuzzy set with a bounded support, and therefore function f is such that

$$
\begin{cases}
f(0) = 1 \\
d(z, Z_{A_{id}}) \geq d_{\max} \implies f[d(z, Z_{A_{id}})] = 0
\end{cases}
\tag{4.3}
$$

Any function with these characteristics may be used, including functions with different branches for both sides of the core of the fuzzy set, generating asymmetric fuzzy sets. (Figure 4.2 shows an asymmetric fuzzy set.) Linear and sinusoidal functions are the most used.

Another source of uncertainty in the definition of the attributes characterizing the GEs occurs when there are several versions of the set Z_A that defines the attribute [46]. This type of uncertainty occurs, for example, when the interval Z_A is determined by experts and several experts have different opinions about its amplitude, or when Z_A is estimated using several methods and different methods determine different intervals, such as in the supervised classification of multispectral remote sensing images, where different classification methods produce different results.

Therefore, as illustrated in Figure 4.5, for some z values, the classification is ambiguous, since, depending on the version used, it may belong or not to Z_A.

To model this type of uncertainty, as in the previous case, attribute A may be represented by a fuzzy set $\widetilde{Z_A}$. Although, in this case, the interpretation of grades of membership as random sets (see [33] or [92]) will be used to compute the grades of membership of the z values to $\widetilde{Z_A}$ ([46]).

Let us consider the random set $\mathcal{A} = \{(Z_{A_j}, m_j) : j = 1, \ldots, n\}$, where the sets Z_{A_j} are the several versions of the set Z_A, and the values m_j are weights assigned to the sets such that

$$
\sum_{j=1}^{n} m_j = 1.
$$

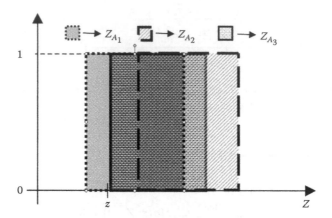

FIGURE 4.5 Three versions of Z_A: Z_{A_1}, Z_{A_2}, and Z_{A_3}. The inclusion of z in set Z_A is then ambiguous.

Then, according to the procedure to compute degrees of membership based on random sets, the membership function of every value z to the fuzzy set $\widetilde{Z_A}$ is given by

$$\mu_{\widetilde{Z_A}}(z) = \sum_{z \in Z_{A_j}} m_j \tag{4.4}$$

4.2.1.2 Positioning an Attribute

In the second phase of the attribute-location approach to build GEs, the geographical location of the attribute chosen in the first phase is determined.

In the cases presented in the previous section where the attribute definition is based on measurable quantities and the uncertainty of the attribute definition is modeled using fuzzy sets, for each location (x_i, y_i) where the measurable quantities Z are known, that is, where a value z_i is known, the uncertainty in the attribute space may be transposed to the geographical space considering

$$\mu_{E_A}(x_i, y_i) = \mu_{\widetilde{Z_A}}(z_i) \tag{4.5}$$

Therefore, whenever z values are known for a tessellation or in every point of the geographical space (when the geographical space is considered continuous), for each location (x_i, y_i) of the geographical space, a degree of membership to the FGE E_A, characterized by attribute A, is obtained.

When no uncertainty is associated to the attribute definition, several sources of errors and uncertainty may be considered in the next step:

- The predefined attribute may be difficult to identify, and therefore it will be difficult to determine its location on the geographical space. For example, the attribute is a certain land cover class, and in some regions, the available data leave doubts about which land cover class is present.

- The attribute definition is not accurate enough, raising doubts about whether entities with some characteristics belong or not to that attribute. For example, if the attribute is buildings, what is exactly considered to be a building? Is a hut a building?
- There is not an abrupt transition between the predefined attribute and other attributes, making the identification of its exact location difficult. For example, if the attribute is forest, it may be difficult to identify the limits of the forest, since there may be a gradual transition between the forest and the surrounding regions.
- The geographical location of the attribute changes with time, and therefore it is difficult to determine its position, such as the coastline or dunes.
- There are errors associated to the measurements made to determine the geographical location of the attributes, for example, observation errors in topographic measurements.
- There are several versions for the entities' position, such as digitalizations made by several operators.

To model the uncertainty or errors regarding the positioning of an attribute, two cases may be considered. One is when the attribute is defined using only a concept such as buildings, forest, or rivers, and the other case is when the attribute definition is based on measurable quantities.

In the first case, the difficulty to position the attributes on the ground will depend upon the details given on its definition and upon the heterogeneity of those attributes on the region under study. The details used in the attribute definition should be such that it is clear what the attribute represents, but note that if too many details are given, the identification of all those details in the ground may complicate the operator's work, since it may be difficult to identify, for example, in an aerial photograph, if a hut is made of wood or brick. So, the attribute definition should be adapted to the methods and sources of information available to identify the entities.

Whenever the operator has some difficulties in the classification, he may always assign a degree of uncertainty to the entity. For example, if there is some uncertainty whether a certain entity should be considered a building or not, a degree of uncertainty may be assigned to it. These degrees of uncertainty are subjective and only indicative that some difficulties in the classification were found. They are assigned to the entity as a whole, since it is an indivisible object. Note that, in this case, the grade of membership represents uncertainty in the attribute that should be assigned to all the region. The outcome of this process is then a GE with a constant grade of membership to an attribute. These grades of membership translate degrees of membership to the attribute defining the GE and not uncertainty on the geographical space. They result from lack of data to assign the correct attribute or lack of attribute definition.

In other situations, the uncertainty is not in the identification of the attribute corresponding to a certain GE, but in the identification of its exact location. In this case, a fuzzy set may be used to express the entities' location in the geographical space. For example, if the attribute is forest and the operator has some difficulties in the identification of its limits, he may define a core where the attribute forest applies for sure and then a surrounding region where there is some transition between forest

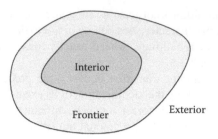

FIGURE 4.6 Egg-yolk approach to represent GEs.

and nonforest. This corresponds to the identification of the core of the fuzzy set and its support, and therefore all the region between the support and the core belongs to the uncertainty region. If it is possible to differentiate further inside the uncertainty region, then α-levels may be identified, if not, only an uncertainty region may be used. This process generates FGEs, which correspond to the "egg-yolk" approach used, for example, in [20] or [38]. This approach considers that the GEs are formed by three regions: the interior, the frontier, and the exterior, where the frontier is represented not by lines, but by a region with any dimension or shape (see Figure 4.6), and that may be considered homogeneous or heterogeneous. The "egg-yolk" representation is a simplified representation of FGEs and is convenient when the GEs are to be represented using the vector data structure and to establish neighborhood relations between GEs with uncertain or fuzzy geographical position.

It is important to stress that, in any of these cases, the grades of membership are subjective and translate human reasoning.

When the attribute definition is based on measurable quantities and no uncertainty in the attribute definition is considered, the attributes are defined by intervals Z_A. The attributes positioning on the geographical space corresponds to determine whether the values z_i measured at location (x_i, y_i) belong or not to interval Z_A. Two sources of uncertainty may be identified in this case:

- There are errors affecting measured values z_i.
- There are several measurements of z_i made at different times.

When there are errors affecting values z_i, they influence the attribute assigned to that location, and therefore the geographical extent of the GEs. When these errors are modeled using a probability distribution function, a fuzzy set, a fuzzy number, or only an interval, the influence of these errors on the classification may be modeled using a method similar to the method proposed by Hisdal to simulate human language (see [55]), where the interpretation of the grades of membership is based on likelihoods. Denoting by $EF(z_i)$ the function modeling the error in value z_i, when $EF(z_i)$ is a probability distribution function, the degree of membership of (x_i, y_i) to the FGEs characterized by attribute A, defined by $Z_A = [\underline{z}, \overline{z}]$, is given by (see [46]).

$$\mu_{E_A}(x_i, y_i) = \int_{\underline{z}}^{\overline{z}} EF(z_i)\,dz_i \tag{4.6}$$

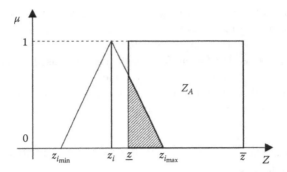

FIGURE 4.7 The area of the shaded region is the grade of membership of location (x_i, y_i) to the FGEs characterized by attribute A, defined by set Z_A, where the uncertainty in the z_i value is modeled with a triangular fuzzy number.

When the error is modeled using fuzzy sets, fuzzy numbers, or intervals, the grades of membership are computed using

$$\mu_{E_A}(x_i, y_i) = \frac{\int_{\underline{z}}^{\bar{z}} EF(z_i) dz_i}{\int_{z_{min}}^{z_{max}} EF(z_i) dz_i} \qquad (4.7)$$

where z_{min} and z_{max} are, respectively, the smaller and larger values that z_i can take according to the error estimation. This value corresponds to the normalized area of the $EF(z_i)$ that is inside set Z_A (see Figure 4.7).

When there are several measurements of z_i, made at different times, these measurements may reflect a changing geographical reality. That is, these measurements reflect a variation over time of what exists in that location. This time variation may be represented using FGEs. In this case, the random set view of fuzzy sets [33] may be used to compute the grades of membership [46]. If n observations are made of the z_i values at all locations, then n versions of the GEs are obtained. The random set $\mathcal{E}_A = \{(E_{Aj}, m_j) | j = 1, \ldots, n\}$ may be considered, where E_{Aj}, with $j = 1, \ldots, n$, represents the n versions of the GE, and the values m_j are weights assigned to each set of observations and therefore to each version, such that $\sum_{j=1}^{n} m_j = 1$. The membership grades of each region to the FGE corresponding to attribute A are given by:

$$\mu_{E_A}(x_i, y_i) = \sum_{(x_i,y_i) \in A_j} m_j \qquad (4.8)$$

4.2.1.3 Identifying Homogeneous Regions

In the location-attribute approach to build GEs, two phases can also be considered:

1. identification of homogeneous regions in the geographical space;
2. determination of the attribute characterizing those regions.

Errors and uncertainty may be found in both phases. The identification of homogeneous regions is subject to error and uncertainty, since it is necessary to determine the degree to which observed characteristics are considered similar, and may therefore be considered to belong to the same region. The resulting regions are supposed to be homogeneous, but a certain degree of heterogeneity will always be present. This heterogeneity may be modeled using fuzzy sets. The most used method to model the heterogeneity is the fuzzy k-means or fuzzy c-means (see, e.g., [16], [56], or [62]). The method enables the clustering of values in k fuzzy classes whose limits are not previously defined. It requires the choice of:

1. the number of classes to be defined;
2. a distance that is used to measure the degree of similarity between the elements of the class;
3. a number $m \in [1, \infty]$, which determines the degree of diversity acceptable within each class (higher values correspond to higher heterogeneity);
4. a positive integer, which determines the degree of similarity required between the elements of the same class, which is usually used as a stop criterion.

This method is usually started choosing a membership grade of each point to each class. Then, the mean of the degrees of membership of the elements that belong to the same class is computed. The next step is to recompute the value of the membership function of each element to each class, minimizing the weighted sum of the distances between each class mean and the degree of membership of the element to each class. With the obtained results, new means are computed for each class and the process is repeated iteratively until the degree of similarity between the elements of each class defined in step 4 is achieved.

Note that, in this case, the degrees of membership to each class are degrees of similarity between each value and an ideal value, which, in each iteration, is the mean of all grades of membership to that class. Note also that, in this method, the membership of each location to the several classes adds up to 1, which means that the classes are not independent of each other [14].

This method has been frequently used to classify several types of phenomena, such as soil types, landforms, and pollution concentration (e.g., [15], [47], [75], and [109]).

4.2.1.4   Choosing an Attribute

The second phase of the classification procedure involves the identification of the attribute corresponding to each of the classes identified in the previous step, which is also subject to uncertainty and error. The fuzzy k-mean clustering method requires the number of classes to be identified *á priori*, which means that some knowledge about the terrain characteristics is necessary in advance. It may, however, happen that the classes obtained with the fuzzy k-mean approach do not describe the terrain adequately, and therefore it may not be easy to choose the attributes that correspond to each of the obtained classes since, on one hand, there may be some heterogeneity in the regions and, on the other hand, there may be doubts in the identification of that attribute. So the previous step may have to be repeated until a realistic classification is obtained.

4.3 PROCESSING FUZZY GEOGRAPHICAL ENTITIES

The use of FGEs in a GIS environment requires operators capable of processing this type of entity. The immediate approach to process FGEs is to convert them into crisp entities and use the usual operators to perform the necessary operations. Since the α-levels of fuzzy sets are crisp sets, the easiest way to convert FGEs to crisp GEs is to substitute the entity by one of its α-levels. With this approach, different versions of the FGE may be obtained according to the needs of each application, choosing different α-levels to represent it. This versatility may be an advantage over the use of common crisp GEs, since different versions of the FGE may be considered according to the needs, but it has also the disadvantage of having different GEs representing the same characteristic in different contexts, which may become confusing if not properly explained in the metadata. This approach, however, does not use the full capabilities of FGEs, since their conversion to crisp entities implies a loss of information regarding their positional uncertainty. Therefore, it is useful to develop operators capable of processing the FGEs without defuzzifying them. These operators work with fuzzy inputs and may have crisp or fuzzy outputs. The operators with fuzzy outputs propagate the fuzziness in the input data to the results of the analysis operations.

Since what characterizes a FGE is the representation of its position in the geographical space, only spatial operators will be considered, that is, operators that process the position of the entity in the geographical space. Some of these operators only generate attributes, which may be crisp or fuzzy, and others generate new entities. Some examples are shown in Table 4.1.

Since the geographical extent of FGEs is represented by fuzzy sets, some operators developed within fuzzy set theory may be used to process FGEs.

4.3.1 COMPLEMENT

The complement of a GE is its exterior, that is, the region that does not belong to it, and therefore is the complement of the set representing the GE. Then, the complement

---

TABLE 4.1
Some Spatial Operations

Operators	Input	Output
Buffer	One entity	New entities
Complement		
Intersection	Several entities	
Reunion		
Area	One entity	New attributes
Perimeter		
Shape		
Distance	Several entities	
Direction		
Neighborhood		

0	0	0	0	0	0.1	0	0	0	0
0	0.1	0.2	0	0.3	0.2	0	0	0	0
0.1	0.3	0.4	0.2	0.5	0.3	0.2	0.1	0	0
0.3	0.4	0.7	0.3	0.5	0.4	0.3	0.2	0.2	0.1
0.2	0.3	1	0.4	0.6	0.5	0.4	0.3	0.2	0
0.1	0.2	1	0.5	0.9	0.6	0.5	0.3	0.1	0
0	0.1	0.6	0.6	1	0.8	0.5	0.2	0	0
0	0	0.3	0.7	1	1	0.8	0.5	0.2	0
0	0	0	0.9	1	1	0.9	0.7	0.3	0.1
0	0	0	0.8	0.9	0.8	0.7	0.4	0.1	0

FIGURE 4.8 Fuzzy geographical entity.

of a FGE is the complement of the fuzzy set representing the entity. The standard complement of a fuzzy set A is the set indicated in Definition 1.6.

Therefore, the standard complement of a FGE E is the FGE CE, such that, for all the space under consideration,

$$\mu_{CE}(x, y) = 1 - \mu_E(x, y)$$

Figure 4.9 shows the standard complement of the FGE shown in Figure 4.8.

4.3.2 UNION AND INTERSECTION

Some authors suggested the use of the standard fuzzy operators of union and intersection to determine the union and intersection of FGEs ([16], [59], [75]).

The standard union of two fuzzy sets A and B is given by Definition 1.4. Then, the union of two FGEs E and F is the FGE, such that, for every point (x, y) of the geographical space (see Figure 4.10),

$$\mu_{E \cup F}(x, y) = \max[\mu_E(x, y), \mu_F(x, y)]$$

1	1	1	1	1	0.9	1	1	1	1
1	0.9	0.8	1	0.7	0.8	1	1	1	1
0.9	0.7	0.6	0.8	0.5	0.7	0.8	0.9	1	1
0.7	0.6	0.3	0.7	0.5	0.6	0.7	0.8	0.8	0.9
0.8	0.7	0	0.6	0.4	0.5	0.6	0.7	0.8	1
0.9	0.8	0	0.5	0.1	0.4	0.5	0.7	0.9	1
1	0.9	0.4	0.4	0	0.2	0.5	0.8	1	1
1	1	0.7	0.3	0	0	0.2	0.5	0.8	1
1	1	1	0.1	0	0	0.1	0.3	0.7	0.9
1	1	1	0.2	0.1	0.2	0.3	0.6	0.9	1

FIGURE 4.9 Complement of fuzzy geographical entity shown in Figure 4.8.

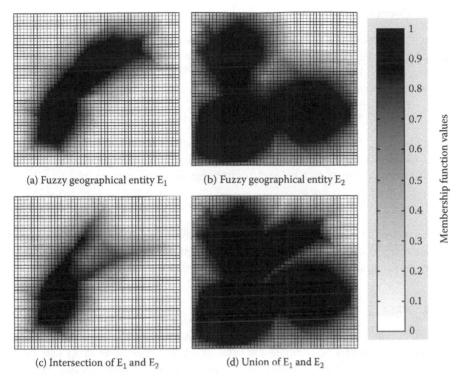

(a) Fuzzy geographical entity E_1 (b) Fuzzy geographical entity E_2

(c) Intersection of E_1 and E_2 (d) Union of E_1 and E_2

FIGURE 4.10 (a) FGE E_1, (b) FGE E_2, (c) intersection of FGEs E_1 and E_2, and (d) union of FGEs E_1 and E_2.

A similar approach may be considered for the intersection. The standard intersection of two fuzzy sets is given by Definition 1.5. Then, the standard intersection of two FGEs E and F is the FGE, that, for every point (x, y) of the geographical space, satisfies (see Figure 4.10),

$$\mu_{E \cap F}(x, y) = \min[\mu_E(x, y), \mu_F(x, y)]$$

4.3.3 BUFFERS GENERATION

Operators to generate buffers around FGEs proposed, for example, by [34] and by [59]. Katinsky proposes that a buffer around FGEs be obtained considering traditional buffers around each α-level of the FGE. In this way, each α-level will generate a region and its degree of membership to the buffer is α (the grade of membership of the α-level that generated it). Fuzzy buffers are also proposed, which introduce another source of uncertainty, since a fixed distance is not used. A more complex approach is presented by [34] and [52], which present several algorithms to generate fuzzy buffers, including iterative methods, global methods, and methods based on the use of graphical software.

4.3.4 DISTANCES AND DIRECTIONS

Altman [1] developed a metric to compute distances between FGEs and directions defined by two FGEs. This approach considers that the distance between two fuzzy sets, represented by discrete points, is a fuzzy set whose elements are the values of the distance between the points of both FGEs. That is,

$$dist(E, F) = \{d_i : d_i = d[(x_1, y_1), (x_2, y_2)], (x_1, y_1) \in E \wedge (x_2, y_2) \in F\}$$

where $d[(x_1, y_1), (x_2, y_2)]$ represents the distance between points (x_1, y_1) and (x_2, y_2). The Euclidean distance can be used, as well as any other distance of interest. The grade of membership of each value d_i to the fuzzy distance is given by

$$\mu_{dist(E,F)}(d_l) = \max_{d_i}\{\min[\mu_E(x_1, y_1), \mu_F(x_2, y_2)]\}$$

The metric developed by Altman [1] for directions is similar to the one developed for distances, replacing the computation of the distance between the points of both sets by the computation of the bearing defined by them.

4.3.5 AREA

The computation of areas of GEs is a basic operation in a GIS. Several approaches have been considered to compute the area of FGEs, generating crisp and fuzzy results. According to Katinsky [59], unambiguous areas of FGEs can only be computed if the FGEs are defuzzified, and no further developments are made. Erwig and Schneider [39] point out that the areas of fuzzy regions are intervals. Their lower limit is the area of the region's core and the upper limit the area of the region's support. Since the use of intervals requires new operators, they proposed considering only minimum and maximum area values corresponding to the lower and upper limits of the intervals.

4.3.5.1 Rosenfeld Area

Rosenfeld [87] proposed an operator to compute the area of a fuzzy set. Since the geographical location of FGEs is represented by a fuzzy set, this operator can be applied to compute the area of FGEs.

The Rosenfeld operator to compute areas (expressed here as the Rosenfeld area RA) considers that the area of a fuzzy set E, formed by n regions r_i, with area $A(r_i)$ and membership function $\mu_E(r_i)$ is given by:

$$RA(E) = \sum_{i=1}^{n} \mu_E(r_i)A(r_i)$$

This operator returns a positive real number and therefore generates a crisp value for the area.

Considering the standard fuzzy intersection, union, and complement operators and establishing a comparison with the three following well-known properties of areas of crisp regions:

1. $\forall E \quad Area(E) \geq 0$
2. $\forall E, F \quad Area(E \cup F) = Area(E) + Area(F) - Area(E \cap F)$
3. $\forall E, F : F \subseteq E \quad Area(E - F) = Area(E) - Area(F)$

where $E - F = E \cap CF$, denoting CF the complement of set F, the Rosenfeld area has the properties listed below. Proofs and further explanations of other characteristics of the Rosenfeld area can be found in [44].

Property RA1: For any FGE E, $RA(E) \geq 0$.
Property RA2: Let E and F be two FGEs, then

$$RA(E \cup F) = RA(E) + RA(F) - RA(E \cap F)$$

Property RA3: If $E \cap F = \phi$, that is, $\nexists r_i : \mu_E(r_i) > 0 \wedge \mu_F(r_i) > 0$, then

$$RA(E \cup F) = RA(E) + RA(F) \qquad (4.9)$$

Property RA4: Let E and F be two FGEs such that $F \subseteq E$. If $\forall r_i \in support(F)$, $\mu_E(r_i) = 1$, that is, $support(F) \subseteq core(E)$, then

$$RA(E - F) = RA(E) - RA(F)$$

Property RA5: For any FGEs E and F such that $F \subseteq E$,

$$RA(E - F) \geq RA(E) - RA(F) \qquad (4.10)$$

As shown in the previous properties, some care must be taken when making operations with Rosenfeld areas, since the Rosenfeld area of the difference of two FGEs may not be equal to the difference of the Rosenfeld areas of both entities.

The method proposed by Rosenfeld to compute the area of a FGE considers that the contribution of the area of each elementary region to the total area is proportional to the degree of membership of the elementary region to the entity. Then, this operator is appropriate to determine the area of a FGE, when the degrees of membership represent the proportion of the area of the elementary region occupied by the attribute characterizing the entity. These FGEs may, for example, be obtained converting GEs from the vector data structure to the raster data structure (for more details, see [43]). However, when the degrees of membership to the FGE represent the uncertainty of whether the elementary regions belong to the GE or not, the spatial extension of the GE is not known and, consequently, its area is also not known. In this case, the value of the area of the FGE obtained with the Rosenfeld operator is just an approximate value of the area, and does not give any other information about the other values it may take. For example, consider a FGE that represents the risk of atmospheric pollution with a certain pollutant, where the degrees of membership of each elementary region to the FGE represent the possibility that the region is affected by the pollutant. Usery presents a simple method to build such a membership function using the semantic import approach [95]. Since any region with $\mu > 0$ may belong to the affected zone, the area of this entity may vary between the areas of the support and the core of the fuzzy set representing the FGE. Note that the degrees of membership of interest may vary with the application. If, for example, the objective is to determine the area of all regions that may be affected by the pollutant, or if only the area which has a possibility of being affected larger than 0.4 is of interest, the Rosenfeld area operator is not useful. The Rosenfeld area operator does not give also any information about the possible variation of the area with the different degrees of membership to the entity.

If the degrees of membership translate degrees of similarity to the attribute characterizing the FGE, the area of the GE depends on the degrees of similarity acceptable for a certain application. For example, if the degrees of membership represent degrees of similarity to the attribute "pine forest," based on the percentage of pine trees observed on the terrain and on the percentage of pine trees considered typical of a pine forest, the area of the entity may, once again, take values between the area of the support and the area of the core of the fuzzy set that represents the FGE. If the objective of a certain application is to determine the area of all regions where pine trees can be found, all degrees of membership are acceptable and the wanted area is the area of the support of the FGE. On the opposite direction, if the objective is to determine the area of the regions where there are only pine trees, only the degrees of membership equal to 1 are of interest. All intermediate situations can be considered, and the Rosenfeld area operator is not useful in either of them.

The previous examples show that the information given by the Rosenfeld area operator is limited and insufficient for many applications.

4.3.5.2　Fuzzy Area

To overcome some of the limitations of the previous operator, a new operator was proposed, called the *fuzzy area operator* [44].

The spatial location of a FGE is represented by a fuzzy set, which is uniquely represented by a family of α-levels, for $\alpha \in [0, 1]$. Since an α-level is a crisp set such that $E_\alpha = \{r_i : \mu_E(r_i) \geq \alpha\}$, its area, denoted by $Area(E_\alpha)$, is the sum of the areas of elementary regions r_i, forming a tessellation, that belong to it.

Let us consider the function

$$Area_E : [0, 1] \rightarrow \mathbb{R}_0^+$$

$$\alpha \rightarrow Area_E(\alpha) = Area(E_\alpha) = z$$

and denote by $z_i, i = 1, \ldots, n$, a set of values of $\mathbb{R}_0^+$ such that $\exists \alpha_i : z_i = Area_E(\alpha_i)$, where $0 < \alpha_i < \alpha_{i+1} \leq 1$.

DEFINITION 4.1
The fuzzy area of a FGE is the fuzzy set $FA(E) = \{(z, \mu_{FA(E)}(z))\}$ with $\mu_{FA(E)}(z) : \mathbb{R}_0^+ \rightarrow [0, 1]$ and

$$\mu_{FA(E)}(z) = \begin{cases} \max_{z_i = Area_E(\alpha_i)} \alpha_i & \text{if } \exists i : z = z_i \\ \frac{z - z_k}{z_{k+1} - z_k}(\alpha_{k+1} - \alpha_k) + \alpha_k & \text{if } \nexists i : z = z_i \wedge z \in [Area_E(1), Area_E(0)] \\ 0 & \text{if } z \notin [Area_E(1), Area_E(0)] \end{cases}$$

(4.11)

where $z_k = \max_{z_i \leq z}(z_i)$ and $\alpha_k = \min_{z_k = Area_E(\alpha_i)}(\alpha_i)$.

0	0	0	0	0	0	0	0	0	0
0	0	0	0	0	0	0.1	0	0	0
0	0	0.1	0.2	0	0.3	0.2	0	0	0
0	0.1	0.3	0.4	0.8	0.5	0.3	0.2	0.1	0
0	0.3	0.4	0.7	1	0.6	0.4	0.3	0.2	0
0	0.2	0.3	1	1	0.8	0.5	0.4	0	0
0	0.1	0.2	1	1	0.9	0.6	0.5	0.3	0
0	0	0.1	0.6	0.6	1	0.8	0.5	0.2	0
0	0	0	0.3	0.7	0.8	0.7	0.8	0.1	0
0	0	0	0	0	0	0	0	0	0

FIGURE 4.11 Fuzzy geographical entity E.

To compute the fuzzy area of the FGE represented in Figure 4.11, some α-levels α_i have to be considered and their areas computed. Table 4.2 shows the areas z_i obtained for the α-levels α_i. Figure 4.12 shows a plot of the fuzzy area of E obtained using the values of Table 4.2. A MATLAB routine to compute fuzzy areas of FGEs can be found in the attached CD (routine **fuzzyarea.m**).

The fuzzy area has some properties corresponding to the area properties indicated in the previous section. These properties are listed below. The proofs and some additional explanations about the fuzzy area may be found in [44].

Property FA1: $\forall E$, the support of $FA(E)$ is a subset of $\mathbb{R}_0^+$.
Property FA2: $\forall E$, $\mu_{FA(E)}(z)$ is a decreasing left-continuous function over $z \in [Area_E(1), Area_E(0)]$.

TABLE 4.2
Area of the α-Levels of E

i	α_i	$z_i = Area(E_{\alpha_i})$
1	0.001	68
2	0.1	68
3	0.2	58
4	0.3	47
5	0.4	36
6	0.5	30
7	0.6	23
8	0.7	19
9	0.8	15
10	0.9	11
11	1	7

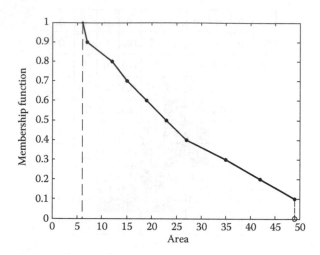

FIGURE 4.12 Fuzzy area of FGE E.

Property FA3: If E is a normal FGE, then $FA(E)$ is a fuzzy number.
Property FA4: If $\not\exists r_i : \mu_E(r_i) > 0 \wedge \mu_F(r_i) > 0$ (that is, $E \cap F = \phi$), then

$$FA(E \cup F) = FA(E) + FA(F) \qquad (4.12)$$

The notation used in the following properties is such that, for a fuzzy set S with α-levels $S_\alpha = [s_\alpha^-, s_\alpha^+]$, $S_\alpha^+ = s_\alpha^+$ and $S_\alpha^- = s_\alpha^-$.
Property FA5: Let E and F be two FGEs, such that $\exists r_i$:

$$\mu_E(r_i) = 1 \wedge \mu_F(r_i) = 1.$$

Then,

$$FA(E \cup F)_\alpha^+ = FA(E)_\alpha^+ + FA(F)_\alpha^+ - FA(E \cap F)_\alpha^+$$
$$\leq [FA(E) + FA(E) - FA(E \cap F)]_\alpha^+ \qquad (4.13)$$

Property FA6: Let E and F be two FGEs, such that $\exists r_i : \mu_E(r_i) = 1 \wedge \mu_F(r_i) = 1$. Then,

$$FA(E \cup F)_\alpha^- = FA(E)_\alpha^- + FA(F)_\alpha^- - FA(E \cap F)_\alpha^-$$
$$\geq [FA(E) + FA(E) - FA(E \cap F)]_\alpha^- \qquad (4.14)$$

Note that properties *FA5* and *FA6* show that

$$FA(E \cup F) \subseteq FA(E) + FA(F) - FA(E \cap F) \qquad (4.15)$$

Property FA7: If E and F are FGEs such that $F \subseteq E$, then

$$FA(E - F)_\alpha^+ \leq [FA(E) - FA(F)]_\alpha^+ \qquad (4.16)$$

Property FA8: If E and F are FGEs such that $F \subseteq core(E)$ and $E - F$ is a normal FGE, then

$$FA(E - F)_\alpha^- = \min[FA(E) - FA(F)]_\alpha^- = FA(E)_1^- - FA(F)_{0+}^+ \qquad (4.17)$$

where $FA(F)_{0+}^+$ represents the largest value of the support of $FA(F)$.

As shown above, the fuzzy area operator generates a fuzzy number, but, even though some properties similar to the area of classic regions hold, in some cases, care must be taken when making operations with the fuzzy areas, since the results of the operations (mainly the difference) may have a behavior different from the expected.

4.3.6 PERIMETER

The perimeter computation of GEs is also a basic operator of a GIS.

4.3.6.1 Rosenfeld Perimeter

Rosenfeld [87] proposed an operator to compute the perimeter of a fuzzy set formed by a finite set of contiguous and homogeneous regions. This operator, herein designated by Rosenfeld perimeter (*RP*), may be applied to FGE represented by a tessellation. The Rosenfeld perimeter is given by:

$$RP(E) = \sum_{\substack{i,j=1 \\ i<j}}^{n} \sum_{k=1}^{n_{ij}} |\mu_E(r_i) - \mu_E(r_j)| l(a_{ijk}) \tag{4.18}$$

where n is the number of elementary regions forming the FGE, n_{ij} is the number of arcs separating regions r_i and r_j and $l(a_{ijk})$ is the length of k arc a_{ijk} of contact between regions r_i and r_j.

This operator considers that each arc separating contiguous elementary regions has a degree of belonging to the perimeter equal to the difference of the membership grades associated to the elementary regions that it separates. Therefore, as for the Rosenfeld area operator, the Rosenfeld perimeter operator generates an approximate real value for the perimeter of the FGE, giving no other information about the other values it can take, or about its variability with the grades of membership to the FGE. To overcome these limitations a fuzzy perimeter operator was proposed.

4.3.6.2 Fuzzy Perimeter

A level cut of a FGE E is a classical set. Let $P(E_\alpha)$ be the perimeter of the α-level cut of the fuzzy set representing the FGE. Let us now consider for each FGE E a function P_E such that

$$P_E : [0, 1] \longrightarrow \mathbb{R}_0^+$$
$$\alpha \longrightarrow P_E(\alpha) = P(E_\alpha) = p \tag{4.19}$$

and denote by p_i, $i = 1, \ldots, n$, a set of values of $\mathbb{R}_0^+$ such that $\exists \alpha_i : p_i = P_E(\alpha_i)$, where $0 < \alpha_i < \alpha_{i+1} < 1$.

DEFINITION 4.2
The fuzzy perimeter of a FGE E is the fuzzy set $FP(E) = \{(z, \mu_{FP(E)}(z))\}$ where

$$\mu_{FP(E)} : \mathbb{R}_0^+ \longrightarrow [0, 1]$$
$$p \longrightarrow \mu_{FP(E)}(p) \tag{4.20}$$

TABLE 4.3
Perimeter of the α-Levels of E

i	α_i	$p_i = P(E_{\alpha_i})$
1	0.001	36
2	0.1	36
3	0.2	34
4	0.3	30
5	0.4	24
6	0.5	22
7	0.6	22
8	0.7	24
9	0.8	24
10	0.9	14
11	1	14

and

$$
\mu_{FP(E)}(p) = \begin{cases} \max\limits_{\substack{p_k, p_{k+1}: \\ (p_k \le p \le p_{k+1} \vee p_{k+1} \le p \le p_k) \\ \wedge p_k \ne p_{k+1}}} \left(\frac{p - p_k}{p_{k+1} - p_k}(\alpha_{k+1} - \alpha_k) + \alpha_k \right) & if \quad p \in [\min p_i, \max p_i] \\ \\ 0 & if \ p \notin [\min p_i, \max p_i] \end{cases} \tag{4.21}
$$

The perimeter values p_i corresponding to the α-levels α_i of the FGE represented in Figure 4.12 are shown in Table 4.3. Figure 4.13 shows the plot of the p_i values shown

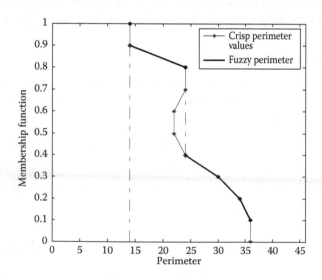

FIGURE 4.13 Fuzzy perimeter of fuzzy geographical entity E.

in Table 4.3 and the fuzzy perimeter of FGE E. A MATLAB routine to compute fuzzy perimeters of FGEs can be found in the attached CD (routine **fuzzyperimeter.m**).

Note that the perimeter-levels 0.5 and 0.6 of α are smaller than the perimeter values of α-level 0.7 (which has a larger area, as can be seen in Table 4.2). Therefore, the fuzzy perimeter is not built in the same way the fuzzy area is because, otherwise in some cases, the output would not be a fuzzy set.

The fuzzy perimeter operator satisfies the following properties. Proofs can be found in [45].

Property FP1: For all FGE E, the support of $PF(E)$ is a subset of $\mathbb{R}_0^+$.
Property FP2: If E is a normal FGE, then $PF(E)$ is a fuzzy number.

Such as with the fuzzy area operator, the support of the fuzzy perimeter shows the set of values the perimeter of the FGE can take and each α-level cut of the fuzzy perimeter is the set of all values the perimeter can take when level cuts of the FGE corresponding to values larger than or equal to α are considered. Therefore, the fuzzy perimeter operator, as the fuzzy area operator, incorporates more information than the crisp operators.

Since the fuzzy perimeter operator generates a fuzzy number, it is possible to perform operations with it, such as the addition of perimeters or the computation of the shape of a FGE, where a fuzzy area may also be used. For more details see [45].

5 Surface Modeling

Jorge Santos

CONTENTS

5.1 INTRODUCTION — DATA UNCERTAINTY IN SURFACE MODELS

To model a continuous geographic phenomenon, first a set of data points needs to be collected from the geographical space. There are two stages in this process: sampling and measurement. Sampling refers to the distribution of measurement places, in order to have a representative sample data set of the phenomena. In this context, the uncertainty arises not only from the inaccuracy of measurements but also from the lack of representativity of sample points and from the choice of the model to be used. The uncertainty will be included in membership functions of fuzzy numbers, which will be used to express attribute values (see Figure 5.1). Therefore, the sample set will by given by

$$f(x_i, \tilde{z}_i), \quad i = 1, 2, \ldots, N, \tag{5.1}$$

where $f(x_i) \in R^n$ is the position of sample $\tilde{z}_i \in F(R)$.

Mathematical interpolators will be used as models for continuous geographic phenomena. The first condition for those interpolators is that they should assume the sampled values in the sampled sites.

$$\tilde{f}(x_i) = \tilde{z}_i, \quad i = 1, 2, \ldots, N. \tag{5.2}$$

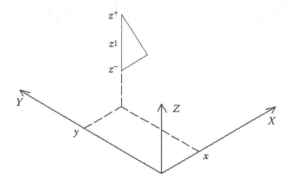

FIGURE 5.1 Triangular fuzzy sample.

Therefore, when sample values are fuzzy numbers, we will have fuzzy interpolators. The fuzzy interpolator can be defined by its α-levels

$$[f(x)]_\alpha = [f_\alpha^-(\mathbf{x}), f_\alpha^+(\mathbf{x})], \quad \alpha \in [0, 1], \tag{5.3}$$

with,

$$\alpha < \alpha' \Rightarrow [f(x)]_{\alpha'} \subset [f(x))]_\alpha, \quad \alpha, \alpha' \in [0, 1]. \tag{5.4}$$

The type of function to be used will have the general formula

$$\tilde{f}(x) = \sum_{j=1}^{n} \tilde{\zeta}_j \phi_j(x), \tag{5.5}$$

where $\phi_j(x)$ are the basis functions, which set the interpolator type. Using α-levels, we get

$$[f(x)]_\alpha = \sum_{j=1}^{n} [\zeta_j]_\alpha \phi_j(x), \quad \alpha \in [0, 1], \tag{5.6}$$

and, expressing intervals in radius/midpoint format, we have

$$[f(x)]_\alpha = \sum_{j=1}^{n} \check{\zeta}_{j,\alpha} \phi_j(x) + \sum_{j=1}^{n} \rho_{j,\alpha} |\phi_j(x)| [-1, 1], \quad \alpha \in [0, 1], \tag{5.7}$$

where $\check{\zeta}_{j,\alpha}$ and $\rho_{j,\alpha}$ are, respectively, the midpoint and the radius of the interval $[\zeta_j]_\alpha$.

Another important condition is that the surfaces generated by those interpolators should behave in the same way as the phenomena that they model. For example, smoothness should be similar. There are several well-known crisp interpolators usually used in geographic modeling. The generalization from crisp to fuzzy interpolators will be done in the way to keep the same properties, namely, smoothness. Consequently, we introduce the notion of *consistent fuzzy interpolator*, which is a fuzzy interpolator whose α-levels' limits are still functions of the same type of the original crisp ones. We will see that this additional condition will force us to relax the conditions in Equation (5.2) to

$$\tilde{f}(x_i) \supseteq \tilde{z}_i, \quad i = 1, 2, \ldots, N, \tag{5.8}$$

or

$$[f(x)]_\alpha = \sum_{j=1}^{n} \check{\xi}_{j,\alpha}\phi_j(x) + \sum_{j=1}^{n} \rho_{j,\alpha}\phi_j(x)[-1, 1] \supseteq [z_i]_\alpha, \quad \alpha \in [0, 1], \qquad (5.9)$$

where the radius $\rho_{j,\alpha}$ should be minimal.

5.2 UNIVARIATE CASE

In the univariate case, we will represent the sampling set by

$\{(x_i, \tilde{z}_i), i = 1, 2, \ldots, N\}$, with the partition $a = x_1 < x_2 < \cdots < x_N = b$.

The problem here is to find a mathematical function $\tilde{f}$ that will model the geographical phenomena within the interval $[a, b] \subset \mathbb{R}$, following the N conditions

$$\tilde{f}(x_i) = \tilde{z}_i, \quad i = 1, 2, \ldots, N$$

or, at least,

$$\tilde{f}(x_i) \supseteq \tilde{z}_i, \quad i = 1, 2, \ldots, N.$$

The type of function to be used will have the general formula

$$\tilde{f}(x) = \sum_{j=1}^{N} \tilde{\xi}_j \phi_j(x),$$

or, using α-levels,

$$[f(x)]_\alpha = \sum_{j=1}^{N} [\zeta_j]_\alpha \phi_j(x), \quad \alpha \in [0, 1],$$

and, expressing intervals in radius/midpoint format,

$$[f(x)]_\alpha = \sum_{j=1}^{N} \check{\xi}_{j,\alpha}\phi_j(x) + \sum_{j=1}^{N} \rho_{j,\alpha}|\phi_j(x)|[-1, 1], \quad \alpha \in [0, 1]. \qquad (5.10)$$

Let us start by the particular example of Lagrange interpolation polynomial, where the basis functions are

$$\phi_j(x) = L_j(x) \equiv \prod_{\substack{i=1 \\ i \neq j}}^{N} \frac{(x - x_i)}{(x_j - x_i)}, \quad j = 1, 2, \ldots, N. \qquad (5.11)$$

In a first approach, the α-levels of Lagrange fuzzy interpolator will be given by

$$[p(x)]_\alpha = \sum_{j=1}^{N} \check{z}_{j,\alpha}L_j(x) + \sum_{j=1}^{N} r_{j,\alpha}|L_j(x)|[-1, 1], \quad \alpha \in [0, 1], \qquad (5.12)$$

where $\check{z}_{j,\alpha}$ and $r_{j,\alpha}$ are, respectively, the midpoint and the radius of a sample α-level.

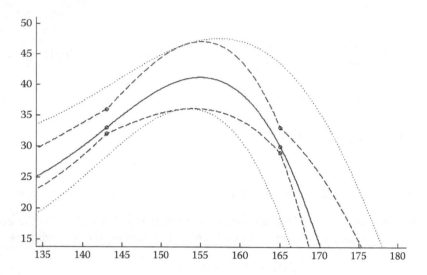

FIGURE 5.2 Triangular fuzzy Lagrange polynomial (support limits in dash, modal curve in continuous, and consistent support limits in dot).

Originally, Lagrange polynomial is infinitely differentiable. However, the fuzzy generalization in Equation (5.12) does not have that property, since the α-levels' limits are

$$p_\alpha^-(x) = \sum_{j=1}^{N} \check{z}_{j,\alpha} L_j(x) - \sum_{j=1}^{N} r_{j,\alpha} |L_j(x)| \text{ and } p_\alpha^+(x)$$

$$= \sum_{j=1}^{N} \check{z}_{j,\alpha} L_j(x) + \sum_{j=1}^{N} r_{j,\alpha} |L_j(x)|. \tag{5.13}$$

The presence of absolute values $|L_j(x)|$ generates nondifferentiable functions (see Figure 5.2). This leads to an inconsistent fuzzy interpolator.

To achieve consistency, a Lagrange polynomial with the same degree should approximate the α-levels' limits given by Equation (5.13). To solve this problem, we have to find the consistent interpolator $\tilde{f}$ with α-levels

$$[f(x)]_\alpha = \sum_{j=1}^{N} \check{\xi}_{j,\alpha} L_j(x) + \sum_{j=1}^{N} \rho_{j,\alpha} L_j(x)[-1, 1], \quad \alpha \in [0, 1], \tag{5.14}$$

such that

$$[p(x)]_\alpha \subseteq [f(x)]_\alpha = \sum_{j=1}^{N} \check{\xi}_{j,\alpha} L_j(x) + \sum_{j=1}^{N} \rho_{j,\alpha} L_j(x)[-1, 1], \quad \alpha \in [0, 1], \tag{5.15}$$

where the radii $\rho_{j,\alpha}$ are minimal. This generates a constrained optimization problem

$$\min_{\rho_\alpha} \|\rho_\alpha\|, \text{ subject to}$$

$$\rho_\alpha^t L(x) - r_\alpha^t |L(x)| \geq 0, \quad \forall x \in [a, b], \alpha \in [0, 1], \tag{5.16}$$

where $\rho_\alpha, r_\alpha, L(x)$, and $|L(x)|$ are vectors with components $\rho_{j,\alpha}, r_{j,\alpha}, L_j(x)$ and $|L_j(x)|$, respectively.

The above optimization problem has an infinite number of conditions (semi-infinite programming). A simple approach to this is to subdivide the interval $[a, b]$, approximating the infinite conditions by a finite number.

Other known interpolators can be fuzzified using the same approach. We will see now the case of other popular interpolators like splines. Linear splines are a simple case of those interpolators. The fuzzy linear splines $\tilde{s}_1$ will be given by its following α-levels

$$[s_1(x)]_\alpha = \sum_{j=1}^N \check{z}_{j,\alpha} s_{1,j}(x) + \sum_{j=1}^N r_{j,\alpha} \left|s_{1,j}(x)\right| [-1, 1], \quad \alpha \in [0, 1], \qquad (5.17)$$

where

$$s_{1,j}(x) = \frac{(x_{j+1} - x)}{(x_{j+1} - x_j)}, \quad \text{and} \quad s_{1,j+1}(x) = \frac{(x - x_j)}{(x_{j+1} - x_j)}, \quad x \in [x_j, x_{j+1}].$$

Since $x_j < x_{j+1}$, we have always $\left|s_{1,j}(x)\right| = s_{1,j}(x)$, $\left|s_{1,j+1}(x)\right| = s_{1,j+1}(x)$, and

$$[s_1(x)]_\alpha = \sum_{j=1}^N \check{z}_{j,\alpha} s_{1,j}(x) + \sum_{j=1}^N r_{j,\alpha} s_{1,j}(x)[-1, 1], \quad \alpha \in [0, 1]. \qquad (5.18)$$

Consequently, fuzzy linear splines are inherently consistent (see Figure 5.3).

The most used splines are the cubic ones, which can be given by

$$s_3(x) = \sum_{j=1}^N z_j s_{3,j}(x),$$

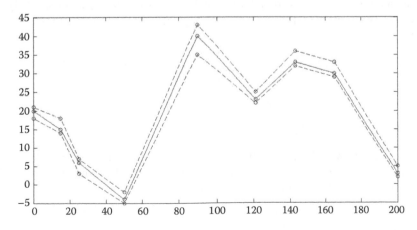

FIGURE 5.3 Triangular fuzzy linear spline (support limits in dash and modal curve in continuous).

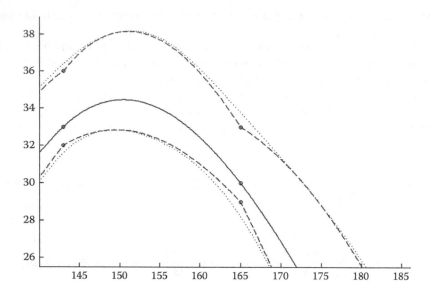

FIGURE 5.4 Triangular fuzzy cubic spline (support limits in dash, modal curve in continuous, and consistent support limits in dot).

where the base functions $s_{3,j}$ are cubic splines interpolating the data points $\{(x_j, \delta_{i,j}),$ $j = 1, 2, \ldots, N\}$, with $\delta_{i.j} = \begin{cases} 1, i = j \\ 0, i \neq j \end{cases}, i = 1, 2, \ldots, N.$ Following the same procedure, a first approach for fuzzy cubic splines will have the α-levels

$$[s_3(x)]_\alpha = \sum_{j=1}^{N} \check{z}_{j,\alpha} s_{3,j}(x) + \sum_{j=1}^{N} r_{j,\alpha} |s_{3,j}(x)|[-1, 1], \quad \alpha \in [0, 1].$$

As in the Lagrange polynomial, the consistency is lost here (see Figure 5.4). To achieve consistency, we follow the same strategy of Lagrange case, getting

$$[f(x)]_\alpha = \sum_{j=1}^{N} \check{z}_{j,\alpha} s_{3,j}(x) + \sum_{j=1}^{N} \zeta_{j,\alpha} s_{3,j}(x)[-1, 1], \quad \alpha \in [0, 1].$$

5.3 BIVARIATE CASE

The natural extension from the univariate to the bivariate case is to use the tensor product, but data have to be arranged in a grid format. However, in most of the practical problems, the sampling set is given by

$$\{(x_i, y_i, \tilde{z}_i), i = 1, 2, \ldots, N\}, \quad \text{in a region } D \subset R^2,$$

where the positions (x_i, y_i) have nongrid distribution. Therefore, first we will study the general case of irregular-spaced data.

The problem now is to find a mathematical function $\tilde{f}$ that will model the geographical phenomena, following the N conditions

$$\tilde{f}(x_i, y_i) = \tilde{z}_i, \quad i = 1, 2, \ldots, N$$

or, at least,

$$\tilde{f}(x_i, y_i) \supseteq \tilde{z}_i, \quad i = 1, 2, \ldots, N.$$

The type of function to be used will have the general formula

$$\tilde{f}(x, y) = \sum_{j=1}^{N} \tilde{\zeta}_j \phi_j(x, y),$$

with α-levels given by

$$[f(x, y)]_\alpha = \sum_{j=1}^{N} \check{z}_{j,\alpha} \phi_j(x, y) + \sum_{j=1}^{N} \zeta_{j,\alpha} |\phi_j(x, y)|[-1, 1], \quad \alpha \in [0, 1]. \quad (5.19)$$

5.3.1 NON-GRIDDED DATA

In this subsection, the data will be considered spatially irregularly distributed over a region (see Figure 5.5). There are several interpolators able to solve this type of problem. These interpolators are usually local, depending only on the existing data in a neighborhood of the interpolation position. The well-known Delaunay triangulation is usually used to find these neighborhoods, as we will see later on.

5.3.1.1 Fuzzy TIN Interpolator

As the name suggests, the triangular irregular network (TIN) interpolator is defined over a triangulation of data positions. Inside every triangle T_k, the interpolator is defined by a plane passing through the triangle's vertex, which is given by

$$f(x, y) = a_k x + b_k y + c_k, (x, y) \in T_k, \quad k = 1, 2, \ldots, N_t, \quad (5.20)$$

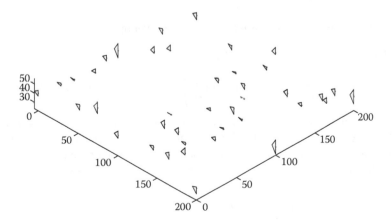

FIGURE 5.5 Example of a fuzzy triangular nongridded data set.

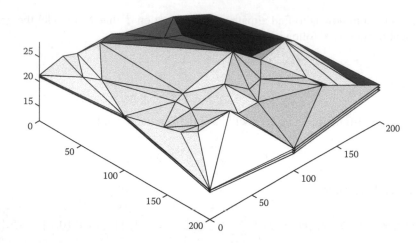

FIGURE 5.6 Fuzzy TIN surface for fuzzy triangular data in Figure 5.5.

where N_t is the number of triangles and a_k, b_k, c_k are the solution of the following linear system of equations

$$A_k x_k = z_k, \text{with} A_k = \begin{bmatrix} x_{k,1} & y_{k,1} & 1 \\ x_{k,2} & y_{k,2} & 1 \\ x_{k,3} & y_{k,3} & 1 \end{bmatrix}, \quad x_k = \begin{bmatrix} a_k \\ b_k \\ c_k \end{bmatrix}, \quad z_k = \begin{bmatrix} z_{k,1} \\ z_{k,2} \\ z_{k,3} \end{bmatrix}. \quad (5.21)$$

Therefore, the resulting surface is a set of contiguous, nonoverlapping triangular facets.

The fuzzy TIN interpolator $\tilde{f}(x, y)$ is defined by its α-levels $[f(x, y)]_\alpha$ for which limits $f_\alpha^-(x, y)$ and $f_\alpha^+(x, y)$ are TINs. For every triangle T_k and for every α-level, the limits $f_\alpha^-(x, y)$ and $f_\alpha^+(x, y)$ are planes determined by solving two linear systems like in Equation (5.21). The only difference is in the right members of equation, which are $z_{k,\alpha}^- = [z_{k,\alpha,1}^- z_{k,\alpha,2}^- z_{k,\alpha,3}^-]^t$ and $z_{k,\alpha}^+ = [z_{k,\alpha,1}^+ z_{k,\alpha,2}^+ z_{k,\alpha,3}^+]^t$, respectively. Consequently, the resulting fuzzy TIN surface will be like in Figure 5.6.

5.3.1.2 Fuzzy Interpolators Based on Weighted Average

Weighted averages can be written in the form

$$f(x, y) = \sum_{j=1}^{N_k} z_j \omega_j(x, y), \text{with} \sum_{j=1}^{N_k} \omega_j(x, y) = 1, \quad (x, y) \in V_k,$$

where V_k is a neighborhood of (x, y) containing points $\{(x_i, y_i) \in V_k, i = 1, 2, \ldots, N_k\}$. The functions ω_j are called weights and usually depend on the position (x, y). A common case is to define those weights based on the inverse of the distances $d_i(x, y)$ between the position (x, y) where the interpolation takes place and the neighborhood points in V_k

$$\omega_j(x, y) = d_j^{-p}(x, y) / \sum_{i=1}^{N_k} d_i^{-p}(x, y),$$

with $d_i(x, y) = \sqrt{(x - x_i)^2 + (y - y_i)^2}$, $p \in \mathbb{N}, i = 1, 2, \ldots, N_k$.

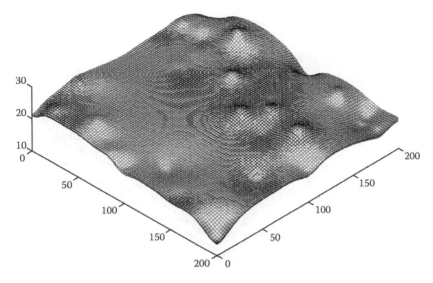

FIGURE 5.7 Fuzzy surface resulting from the fuzzy Shepard method ($p = 2$) for fuzzy triangular data in Figure 5.5.

This choice of weights generates the so-called Shepard method or inverse distance weighting (IDW) of interpolation, which is a simple approach to model geographical phenomena, where usually dependency is inversely proportional to distance (see Figure 5.7). Since the weight functions ω_j are always non-negative, the generalization to fuzzy Shepard method is straightforward. The fuzzy interpolator will be given by

$$\tilde{f}(x, y) = \sum_{j=1}^{N_k} \check{z}_j \omega_j(x, y), \quad (x, y) \in V_k,$$

with α-levels expressed by

$$[f(x, y)]_\alpha = \sum_{j=1}^{N_k} \check{z}_{j,\alpha} \omega_j(x, y)$$

$$+ \sum_{j=1}^{N_k} r_{j,\alpha} \omega_j(x, y)[-1, 1], \quad \alpha \in [0, 1], \quad (x, y) \in V_k. \quad (5.22)$$

5.3.1.3   Fuzzy Interpolators Based on Radial Functions

Weighted functions are an example of radial functions, since they do not depend on the direction but only a distance. Another interpolator based on radial functions is the thin-plate splines (see Figure 5.8). This interpolator is defined by

$$S(x, y) = \sum_{j=1}^{M} a_j \kappa_j(x, y) + a_{M+1} + a_{M+2} x + a_{M+3} y, \quad (5.23)$$

where

$$\kappa_j(x, y) = \frac{1}{8\pi} d_j^2(x, y) \ln[d_j(x, y)], \quad d_j^2(x, y) = (x - x_j)^2 + (y - y_j)^2. \quad (5.24)$$

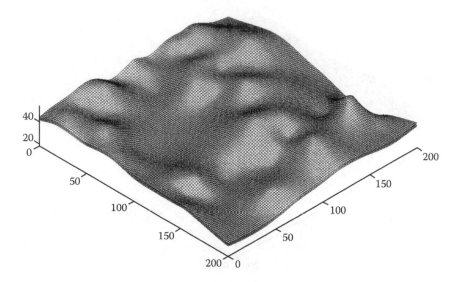

FIGURE 5.8 Fuzzy surface resulting from the fuzzy thin-plate spline interpolator for fuzzy triangular data in Figure 5.5.

The coeficients $a_j (j = 1, 2, \ldots, M + 3)$ are found by solving the linear system of equations

$$Aa = z, \tag{5.25}$$

where

$$
A = \begin{bmatrix} K & P \\ P^t & 0 \end{bmatrix}, \quad a = \begin{bmatrix} a_1 \\ \vdots \\ a_M \\ a_{M+1} \\ a_{M+2} \\ a_{M+3} \end{bmatrix}, \quad z = \begin{bmatrix} z_1 \\ \vdots \\ z_M \\ 0 \\ 0 \\ 0 \end{bmatrix}, \tag{5.26}
$$

with

$$
K = \begin{bmatrix} \kappa_1(x_1, y_1) & \cdots & \kappa_M(x_1, y_1) \\ \vdots & \ddots & \vdots \\ \kappa_1(x_M, y_M) & \cdots & \kappa_M(x_M, y_M) \end{bmatrix}, \quad P = \begin{bmatrix} 1 & x_1 & y_1 \\ \vdots & \vdots & \vdots \\ 1 & x_M & y_M \end{bmatrix}. \tag{5.27}
$$

Therefore, this interpolator can also be written in the form

$$S(x, y) = \sum_{j=1}^{N} z_j \phi_j(x, y), \tag{5.28}$$

where $N = M + 3$ and

$$\phi(x, y) = (A^{-1})^t \kappa(x, y), \tag{5.29}$$

with $\phi(x, y) = [\phi_1(x, y) \cdots \phi_N(x, y)]^t$ and $\kappa(x, y) = [\kappa_1(x, y) \cdots \kappa_M(x, y) \; 1 \; x \; y]^t$.

Three zero-attribute values ($z_{M+1} = z_{M+2} = z_{M+3} = 0$) were added to keep the general interpolation formula in Equation (5.28). Now we can derive the fuzzy version for this interpolator, getting

$$\tilde{S}(x, y) = \sum_{j=1}^{N_k} \tilde{z}_j \phi_j(x, y),$$

with α-levels expressed by

$$[S(x, y)]_\alpha = \sum_{j=1}^{N_k} \check{z}_{j,\alpha} \phi_j(x, y) + \sum_{j=1}^{N_k} r_{j,\alpha} |\phi_j(x, y)|[-1, 1], \quad \alpha \in [0, 1]. \quad (5.30)$$

Here we have, again, the problem of lack of consistency because the basis functions $\phi_j(x, y)$ do not have constant sign. To have consistent fuzzy thin-plate splines, we have to follow the same procedure we did for Lagrange polynomial and cubic splines, getting

$$[f(x, y)]_\alpha = \sum_{j=1}^{N_k} \check{z}_{j,\alpha} \phi_j(x, y) + \sum_{j=1}^{N_k} \zeta_{j,\alpha} \phi_j(x, y)[-1, 1], \quad \alpha \in [0, 1]. \quad (5.31)$$

As it was noticed in the begining of this subsection, these interpolators use most of the time only the neighbor data values, especially when there is a large amount of data to interpolate.

5.3.2 GRIDDED DATA

Now we will consider that the sample data are distributed over a rectangular grid (see Figure 5.9). There are interpolators that can be applied only in this context.

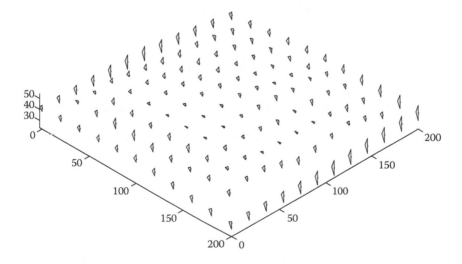

FIGURE 5.9 Example of a fuzzy triangular gridded data set.

Therefore, before applying an interpolator of this type to a set of samples distributed irregularly, first an interpolator for nongridded data has to be used to generate a gridded distribution. The grid is defined by partitions over a rectangular region $D = [a, b] \times [c, d] \subset \mathbb{R}^2$

$$a = x_1 < x_2 < \cdots < x_N = b,$$

$$c = y_1 < y_2 < \cdots < y_M = d.$$

As it was mentioned before, bivariate interpolators can be derived from univariate ones by using the tensor product. Taking the general interpolator formula

$$\tilde{f}(x, y) = \sum_{i=1}^{N} \sum_{j=1}^{M} \tilde{\xi}_{ij} \phi_{ij}(x, y), \tag{5.32}$$

we write the basis functions as a tensor product

$$\phi_{ij}(x, y) = \phi_i(x)\phi_j(y). \tag{5.33}$$

This allows us to build bivariate fuzzy interpolators from univariate ones.

5.3.2.1 Bivariate Lagrange Polynomial

The bivariate Lagrange polynomial is usually written in the form

$$f(x, y) = \sum_{i=1}^{N} \sum_{j=1}^{M} z_{ij} L_{ij}(x, y), \tag{5.34}$$

where the bivariate basis functions are written as a tensor product

$$L_{ij}(x, y) = L_i(x)L_j(y), \tag{5.35}$$

with

$$L_i(x) \equiv \prod_{\substack{k=1 \\ k \neq i}}^{N} \frac{(x - x_k)}{(x_i - x_k)}, \quad i = 1, 2, \ldots, N \text{ and}$$

$$L_j(y) \equiv \prod_{\substack{k=1 \\ k \neq j}}^{M} \frac{(y - y_k)}{(y_j - y_k)}, \quad j = 1, 2, \ldots, M. \tag{5.36}$$

The first approach for fuzzy bivariate Lagrange interpolator will be given by

$$\tilde{f}(x, y) = \sum_{i=1}^{N} \sum_{j=1}^{M} \tilde{z}_{ij} L_{ij}(x, y), \tag{5.37}$$

with the following α-levels

$$[f(x, y)]_\alpha = \sum_{i=1}^{N} \sum_{j=1}^{M} \check{z}_{ij,\alpha} L_{ij}(x, y)$$

$$+ \sum_{i=1}^{N} \sum_{j=1}^{M} r_{ij,\alpha} |L_{ij}(x, y)|[-1, 1], \quad \alpha \in [0, 1]. \tag{5.38}$$

As in the univariate case, consistency is lost here. Therefore, the consistent bivariate fuzzy Lagrange polynomial, defined by the α-levels

$$[f(x, y)]_\alpha = \sum_{i=1}^{N} \sum_{j=1}^{M} \check{z}_{ij,\alpha} L_{ij}(x, y)$$
$$+ \sum_{i=1}^{N} \sum_{j=1}^{M} \zeta_{ij,\alpha} L_{ij}(x, y)[-1, 1], \quad \alpha \in [0, 1], \qquad (5.39)$$

has to be found using a similar procedure.

5.3.2.2 Bilinear Fuzzy Splines

The fuzzy bilinear spline will be given by its α-levels

$$[f(x, y)]_\alpha = \sum_{i=1}^{N} \sum_{j=1}^{M} \check{z}_{ij,\alpha} s_{1,ij}(x, y)$$
$$+ \sum_{i=1}^{N} \sum_{j=1}^{M} r_{ij,\alpha} |s_{1,ij}(x, y)|[-1, 1], \quad \alpha \in [0, 1]. \qquad (5.40)$$

Using again the tensor product, we have

$$s_{1,ij}(x, y) = s_{1,i}(x) s_{1,j}(y), \quad (i = 1, 2, \ldots, N), (j = 1, 2, \ldots, M), \qquad (5.41)$$

where the only nonzero basis functions are

$$s_{1,i}(x) = \frac{(x_{i+1} - x)}{(x_{i+1} - x_i)}, \quad s_{1,i+1}(x) = \frac{(x - x_i)}{(x_{i+1} - x_i)},$$

$$s_{1,j}(y) = \frac{(y_{j+1} - y)}{(y_{j+1} - y_j)} \text{ and } s_{1,j+1}(y) = \frac{(y - y_j)}{(y_{j+1} - y_j)},$$

with $(x, y) \in [x_i, x_{i+1}] \times [y_j, y_{j+1}]$.

Since $x_i < x_{i+1}$ and $y_j < y_{j+1}$, we have always $|s_{1,ij}(x, y)| = s_{1,ij}(x, y)$. Therefore, as in the univariate case, consistency is maintained. The α-levels for bilinear fuzzy splines are

$$[f(x, y)]_\alpha = \sum_{i=1}^{N} \sum_{j=1}^{M} \check{z}_{ij,\alpha} s_{1,ij}(x, y)$$
$$+ \sum_{i=1}^{N} \sum_{j=1}^{M} r_{ij,\alpha} s_{1,ij}(x, y)[-1, 1], \alpha \in [0, 1]. \qquad (5.42)$$

5.3.2.3   Bicubic Fuzzy Splines

Following the same method, in a first approach, the bicubic fuzzy splines have the
following α-levels

$$[f(x, y)]_\alpha = \sum_{i=1}^{N} \sum_{j=1}^{M} \check{z}_{ij,\alpha} s_{3,ij}(x, y)$$

$$+ \sum_{i=1}^{N} \sum_{j=1}^{M} r_{ij,\alpha} \left| s_{3,ij}(x, y) \right| [-1, 1], \alpha \in [0, 1]. \qquad (5.43)$$

The tensor product allows us once more to express the bivariate basis functions using
the univariate ones. The result is

$$s_{3,ij}(x, y) = s_{3,i}(x) s_{3,j}(y), \qquad (5.44)$$

where $s_{3,i}$ and $s_{3,j}$ are cubic splines interpolating the data points $\{(x_i, \delta_{i,j}), i = 1, 2, \ldots, N, \}$ and $\{(y_j, \delta_{i,j}), j = 1, 2, \ldots, M\}$, respectively, with $\delta_{i.j} = \begin{cases} 1, i = j \\ 0, i \neq j \end{cases}$.

As in the Lagrange polynomial, the consistency is lost in this case. Therefore,
the consistent bicubic spline should be redefined by the following α-levels (see
Figure 5.10)

$$[f(x, y)]_\alpha = \sum_{i=1}^{N} \sum_{j=1}^{M} \check{z}_{ij,\alpha} s_{3,ij}(x, y)$$

$$+ \sum_{i=1}^{N} \sum_{j=1}^{M} \zeta_{ij,\alpha} s_{3,ij}(x, y)[-1, 1], \alpha \in [0, 1]. \qquad (5.45)$$

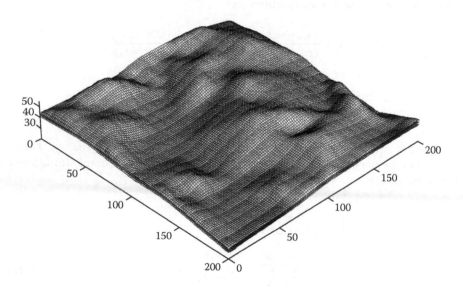

FIGURE 5.10 Fuzzy surface resulting from the fuzzy bicubic spline interpolator for fuzzy
triangular data in Figure 5.9.

As in the other cases, the consistent interpolator has to be found using an approximation procedure as described before.

5.3.3 FUZZY GEOSTATISTIC INTERPOLATOR (KRIGING)

Geostatistics was developed by the French mathematician Georges Matheron in the early sixties, from the seminal work of its inventor, Daniel G. Krige, a South African mining engineer, from whom the name *kriging* is derived [36]. The geostatistic interpolation is often used to model geographical phenomena. Often, the sampled data used in the interpolation have uncertainty that can be expressed by fuzzy numbers ([8], [7], [25]). In this section, the geostatistic tools will be extended to deal with samples expressed by triangular fuzzy numbers in the form

$$\tilde{z} = \tilde{z}(x, y) = (z^-(x, y)/z^1(x, y)/z^+(x, y)), \quad (x, y) \in D \subset \mathrm{R}^2. \tag{5.46}$$

The geographical phenomena will be modeled by a triangular fuzzy random field

$$\{\tilde{Z}(x, y) = (Z^-(x, y)/Z^1(x, y)/Z^+(x, y)), \quad (x, y) \in D\} \tag{5.47}$$

and the sample set of triangular fuzzy numbers $\{\tilde{z}(x_i, y_i) = (z^-(x_i, y_i)/z^1(x_i, y_i)/z^+(x_i, y_i)), i = 1, 2, \ldots, N\}$ will be a set of realizations of the random field.

Using the Hausdorff metric η between the α-levels, a metric d_* can be defined over the set of fuzzy numbers F, given by

$$d_*(\tilde{z}_1, \tilde{z}_2) = \sup_{\alpha \in (0, 1]} \eta([z_1]_\alpha, [z_2]_\alpha). \tag{5.48}$$

For triangular fuzzy numbers, an equivalent metric can be defined by

$$d(\tilde{z}_1, \tilde{z}_2)^2 = (z_1^- - z_2^-)^2 + (z_1^1 - z_2^1)^2 + (z_1^+ - z_2^+)^2. \tag{5.49}$$

Generalizing last metric for any fuzzy number defined by its α-levels, we have

$$d(\tilde{z}_1, \tilde{z}_2)^2 = \sum_\alpha [(z_{1,\alpha}^- - z_{2,\alpha}^-)^2 + (z_{1,\alpha}^+ - z_{2,\alpha}^+)^2].$$

Diamond [25] proved that the expectation $E\tilde{Z}(x, y)$ exists if and only if Ed $(\tilde{Z}(x, y), \tilde{0})^2$, where $\tilde{0} = (0/0/0)$, exists and is a triangular fuzzy number given by

$$E\tilde{Z}(x, y) = (EZ^-(x, y)/EZ^1(x, y)/EZ^+(x, y));$$

Besides the variance $Var\tilde{Z}$ is defined by

$$Var\tilde{Z}(x, y) = Ed(\tilde{Z}(x, y), E\tilde{Z}(x, y))^2 \tag{5.50}$$

and it is a real number.

To express spatial variability, the so-called *stationary covariance* or *covariagram* function $\tilde{C}$, which is defined by the covariance between $\tilde{Z}(x_i, y_i)$ and $\tilde{Z}(x_j, y_j)$, is used.

$$\tilde{C}(x_i - x_j, y_i - y_j) = cov(\tilde{Z}(x_i, y_i), \tilde{Z}(x_j, y_i)).$$

We say that the random variable $\tilde{Z}(x, y)$ is *second order stationary* if the expected value exists and is independent of location (x, y), that is,

$$E\tilde{Z}(x, y) = \tilde{m} = (m^-/m^1/m^+), \quad \forall (x, y) \in D, \tag{5.51}$$

and there exist lower, C^-, modal, C^1, and upper covariance, C^+, that are functions of the lag vector h_{ij} between positions (x_i, y_i) and (x_j, y_j) such that

$$C^-(h_{ij}) = E[Z^-(x_i, y_i)Z^-(x_j, y_j)] - (m^-)^2, \qquad (5.52)$$

$$C^1(h_{ij}) = E[Z^1(x_i, y_i)Z^1(x_j, y_j)] - (m^1)^2, \qquad (5.53)$$

$$C^+(h_{ij}) = E[Z^+(x_i, y_i)Z^+(x_j, y_j)] - (m^+)^2. \qquad (5.54)$$

The assumption of second order stationarity is common in geostatistics. It represents physical homogeneity and the existence of the first two moments, which are independent of location. In practical situations, the observations may show a systematic trend, and it cannot be assumed that the mean is constant. Universal kriging takes this into account, or then a small "moving window" of adjacent points can be used to take a moving average, where the constancy of $E\tilde{Z}(x, y) = \tilde{m}$ is not a bad approximation.

Traditionally, the variogram has been used for modeling spatial variability rather than covariance. It is an alternative to the covariance and is defined as the variance of the increment $Z(x_i, y_i) - Z(x_j, y_j)$. If both exist, they are related by

$$\gamma(h_{ij}) = C(0) - C(h_{ij}). \qquad (5.55)$$

Let $\tilde{\gamma}(h) = (\gamma^-(h)/\gamma^1(h)/\gamma^+(h))$ represent the fuzzy triangular variogram. Assuming the existence of lower, γ^-, modal, γ^1, and upper semivariogram, γ^+, independent of (x, y), the stationarity condition may be expressed by

$$2\gamma^-(h_{ij}) = E[Z^-(x_i, y_i) - Z^-(x_j, y_j)]^2, \qquad (5.56)$$

$$2\gamma^1(h_{ij}) = E[Z^1(x_i, y_i) - Z^1(x_j, y_j)]^2, \qquad (5.57)$$

$$2\gamma^+(h_{ij}) = E[Z^+(x_i, y_i) - Z^+(x_j, y_j)]^2. \qquad (5.58)$$

5.3.3.1 Estimation of Fuzzy Variogram

In crisp kriging, the experimental variogram is given by

$$\hat{\gamma}(h_{ij}) = \frac{\sum_{i=1}^{n(h)}[Z(x_i, y_i) - Z(x_j, y_j)]^2}{2n(h_{ij})} \qquad (5.59)$$

where $n(h)$ is the number of data pairs separated by the same h; under isotropy conditions only, magnitude $\|h\|$ is considered.

If data $\tilde{Z}(x_1), \tilde{Z}(x_2), \ldots, \tilde{Z}(x_n)$ are fuzzy numbers, then fuzzy arithmetic operations [2] can be used to evaluate experimental variogram

$$\hat{\gamma}(h) = \frac{\bigoplus_{i=1}^{n(h)}[(\tilde{Z}(x_i, y_i) \ominus \tilde{Z}(x_j, y_j)) \otimes (\tilde{Z}(x_i, y_i) \ominus \tilde{Z}(x_j, y_j))]}{2n(h_{ij})}. \qquad (5.60)$$

After the experimental variogram has been calculated, a theoretical variogram has to be chosen and fitted to it (see Figures 5.11 and 5.12). The classical method to fit

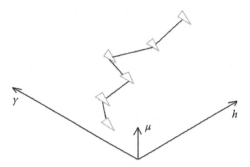

FIGURE 5.11 Fuzzy experimental variogram.

a model γ to a finite set of data (h_i, γ_i), $(i = 1, 2, \ldots, N(h))$ is based on the least squares method. Extending that method to fit a fuzzy model $\tilde{\gamma}$ to fuzzy data $(h_i, \tilde{\gamma}_i)$, we have to minimize

$$F(\tilde{p}_1, \tilde{p}_2, \ldots, \tilde{p}_q) = \sum_{i=1}^{N(h)} d(\tilde{\gamma}_i, \tilde{\gamma}^*(h_i))^2, \tag{5.61}$$

where $(\tilde{p}_1, \tilde{p}_2, \ldots, \tilde{p}_q)$ are the fuzzy parameters of $\tilde{\gamma}^*$.

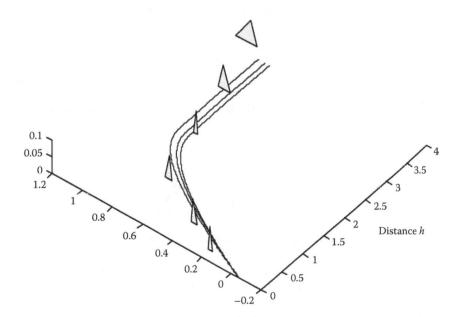

FIGURE 5.12 Fuzzy theoretical variogram fitted to experimental fuzzy values.

A distance d between fuzzy numbers has to be defined. Taking the metric in Equation (5.49), we have

$$F(\tilde{p}_1, \tilde{p}_2, \ldots, \tilde{p}_q) = \sum_{i=1}^{N(h)} \left[(\gamma_i^- - \gamma^-(h_i))^2 + (\gamma_i^1 - \gamma^1(h_i)^2) + (\gamma_i^+ - \gamma^+(h_i))^2 \right].$$

(5.62)

Using, for example, a fuzzy spherical theoretical variogram model

$$\tilde{\gamma}^*(h) = \begin{cases} \tilde{0}, & h = 0 \\ \tilde{C}_0 \oplus \tilde{C}_1 \otimes \left(\frac{3h}{2a} - \frac{h^3}{2a^3} \right), & 0 < h \le a, \\ \tilde{C}_0 \oplus \tilde{C}_1, & h > a \end{cases}$$

(5.63)

where the variogram parameters are the *range* a, the *nugget effect* $\tilde{C}_0$, and the *sill* $\tilde{C}_0 \oplus \tilde{C}_1$. Since a is a distance, it will be considered as a real value. Taking $\tilde{C}_0$ and $\tilde{C}_1$ as triangular fuzzy numbers, we have $\tilde{C}_0 = (C_0^-/C_0^1/C_0^+)$ and $\tilde{C}_1 = (C_1^-/C_1^1/C_1^+)$. So, the variogram can be written as

$$\tilde{\gamma}^*(h) = \begin{cases} (0/0/0), & h = 0 \\ (C_0^-/C_0^1/C_0^+) \oplus (C_1^-/C_1^1/C_1^+) \otimes \left(\frac{3h}{2a} - \frac{h^3}{2a^3} \right), & 0 < h \le a \\ (C_0^-/C_0^1/C_0^+) \oplus (C_1^-/C_1^1/C_1^+), & h > a \end{cases}$$

$$= \begin{cases} (0/0/0), & h = 0 \\ \left(C_0^- + C_1^- \left(\frac{3h}{2a} - \frac{h^3}{2a^3} \right) / C_0^1 + C_1^1 \left(\frac{3h}{2a} - \frac{h^3}{2a^3} \right) / C_0^+ \right. \\ \left. + C_1^+ \left(\frac{3h}{2a} - \frac{h^3}{2a^3} \right) \right), & 0 < h \le a \\ (C_0^- + C_1^-/C_0^1 + C_1^1/C_0^+ + C_1^+), & h > a \end{cases}.$$

(5.64)

Substituting in Equation (5.62), using $\eta_i(a) = \frac{h_i^3}{2a^3} - \frac{3h_i}{2a}$ to simplify, it turns out

$$F(a, \tilde{C}_0, \tilde{C}_1) = \sum_{i=1}^{N(h \le a)} \left[(\gamma_i^- + C_0^- + C_1^- \eta_i(a))^2 + (\gamma_i^1 + C_0^1 + C_1^1 \eta_i(a))^2 \right.$$

$$\left. + (\gamma_i^+ + C_0^+ + C_1^+ \eta_i(a))^2 \right] + \sum_{N(0 < h \le a)}^{N(h > a)} \left[(\gamma_i^- - C_0^- - C_1^-)^2 \right.$$

$$\left. + (\gamma_i^1 - C_0^1 - C_1^1)^2 + (\gamma_i^+ - C_0^+ - C_1^+)^2 \right].$$

(5.65)

Assuming that F is a function of $u = (a, C_0^-, C_0^1, C_0^+, C_1^-, C_1^1, C_1^+)$, we want to find u^* that minimizes F subject to conditions $0 < a$, $0 \le C_0^- C_0^1 \le C_0^+$, and $0 \le C_1^- \le C_1^1 \le C_1^+$. This theoretical variogram will be used to model spatial variability, providing, by Equation (5.55), the covariance values for the kriging process (see Figure 5.13).

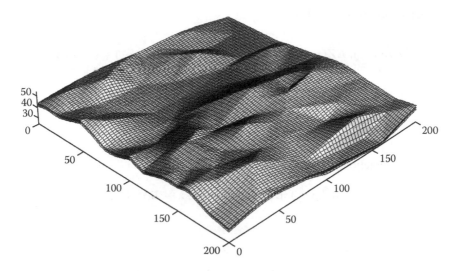

FIGURE 5.13 Fuzzy surface resulting from the fuzzy kriging interpolator for fuzzy triangular data in Figure 5.5.

5.3.3.2 Kriging Fuzzy Data

To estimate the value $\tilde{Z}(x_0, y_0)$ at a position $(x_0, y_0) \in D$, where no observation is known, we may use the fuzzy estimator

$$\hat{Z}(x) = \bigoplus_{i=1}^{n} \lambda_i \odot \tilde{Z}(x_i),$$

where sum $\oplus$ and scalar multiplication $\odot$ are operations defined for fuzzy numbers (see, for example chapter 1).The weights λ_i are to be estimated so that $\hat{Z}(x)$ is unbiased, $E\hat{Z}(x) = E\tilde{Z}(x) = \tilde{m}$ and the variance $Var\hat{Z}(x) = Ed(\hat{Z}(x), E\tilde{Z}(x))^2$ is minimum. From the first condition, we get

$$\sum_{i=1}^{n} \lambda_i = 1. \tag{5.66}$$

To avoid mixed products, we impose also that the weights should be non-negative

$$\lambda_i \geq 0, \quad i = 1, 2, \ldots, n. \tag{5.67}$$

Therefore, we get

$$\hat{Z}(x) = \left(\sum_{i=1}^{n} \lambda_i Z^-(x_i, y_i) / \sum_{i=1}^{n} \lambda_i Z^1(x_i, y_i) / \sum_{i=1}^{n} \lambda_i Z^+(x_i, y_i) \right),$$

and adding the minimum variance condition, we have the following optimization problem to solve

$$\min_{\substack{\sum_{i=1}^{n} \lambda_i = 1, \\ \lambda_i \geq 0, \, i=1,2,\ldots,n}} Ed(\hat{Z}(x, y), E\tilde{Z}(x, y))^2.$$

Diamond([23]) shows that

$$Ed(\hat{Z}(x, y), E\tilde{Z}(x, y))^2 = \sum_{i,j=1}^{n} \lambda_i \lambda_j (C^-(h_{ij}) + C^1(h_{ij}) + C^+(h_{ij}))$$

$$- 2\sum_{i=1}^{n} \lambda_i (C^-(h_{i0}) + C^1(h_{i0}) + C^+(h_{i0}))$$

$$+ C^-(0) + C^1(0) + C^+(0)$$

or, in matrix notation,

$$Ed(\hat{Z}(x, y), E\tilde{Z}(x, y))^2 = \lambda^t C \lambda - 2\lambda^t c + c,$$

with

$$\lambda = [\lambda_i]_{i=1,2,\ldots,n},$$
$$C = [C_{ij}]_{i,j=1,2,\ldots,n}, \quad C_{ij} = C^-(h_{ij}) + C^1(h_{ij}) + C^+(h_{ij}),$$
$$c = [c_i]_{i=1,2,\ldots,n}, \quad c_i = C^-(h_{i0}) + C^1(h_{i0}) + C^+(h_{i0}),$$
$$c = C^-(0) + C^1(0) + C^+(0).$$

Then, the optimization problem to solve in order to get the weights λ_i has the constrained quadratic form

$$\min_{\substack{\sum_{i=1}^{n} \lambda_i = 1, \\ \lambda_i \geq 0, i=1,2,\ldots,n}} \lambda^t C \lambda - 2\lambda^t c + c.$$

Introducing the Lagrange multiplier μ for the condition in Equation (5.66) and multipliers $v_1, v_2, \ldots, v_n$ for the conditions in Equation (5.67), the weights λ_i must satisfy the the Kuhn-Tucker conditions [73], leading to the system

$$\sum_{i=1}^{n} C_{ij}\lambda_i - v_j - \mu = c_j \quad j = 1, 2, \ldots, n$$

$$\sum_{i=1}^{n} \lambda_i = 1$$

$$\sum_{i=1}^{n} v_i \lambda_i = 0$$

$$v_i, \lambda_i \geq 0 \quad i = 1, 2, \ldots, n$$

6 Visualization and Analysis on Surfaces

Giovanni Gallo

CONTENTS

6.1   GENERAL CONSIDERATIONS

Errors and uncertainty are an important issue in any form of scientific communication. It is hence natural that many tools and standard techniques have been developed for 2D visualization of scientific data since the early start of scientific visualization, representing uncertainties of points in 2D as vertical bars, graphing probability density curves and surfaces, and displaying uncertainty side by side. When the data sets are large, complex, and/or multidimensional, the problem becomes more challenging. It is hence not surprising that the visualization research has devoted several efforts to the problem of the visual representation of errors and uncertainty for 3D visualization. Indeed several applications have a crucial need to correctly visualize uncertainty in the 3D case: Display of medical information and GIS are perhaps the two most relevant examples of such applications.

The visualization of uncertainty is of course an imperative when dealing with uncertain phenomena, but uncertainty is present at the core of the visualization problem, and has to be taken into consideration even if the data to represent would be perfectly known. Indeed, visualization of natural phenomena per se is a major source of uncertainty; almost every visualization technique requires data filtering and interpolation. Attention is paid to this relevant issue not too often; the error introduced because of the adoption of some visual technique adds itself to the unavoidable errors and uncertainty introduced by acquisition, modeling, and data transformation. These ideas have been explicitly brought into the focus of the scientific visualization research in the late nineties when a small set of influential papers about this subject was published ([22], [49], [85]). A schematic view of "uncertainty pipeline" in data visualization may be found in [64] and in [85], and may be summarized as follows: Data coming from physical phenomena acquire uncertainty in the measurement phase, in the transformation phase, and in the computation of the visual output for their representation.

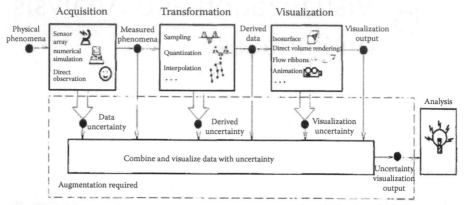

FIGURE 6.1 Uncertainty pipeline from physical phenomena to visualization (From Lodha et al., 1996. With permission).

A good visualization system to support the final visual analysis of the data requires an "augmentation" of the standard techniques to include uncertainty information. This path leads to the idea of "verity representation": Data and uncertainty should be represented together in order to provide to the user a holistic way to perceive them and to draw correct inferences from the pictures. A schematic view of this pipeline is given in Figure 6.1 (from [64]).

It is important to stress that there is no absolute better technique to solve the problem of visualizing uncertainty in 3D data; in his short note, Gershon [49] pointed out that any representation is imperfect and visual representations cannot escape this destiny, and introduced the concept of "acceptable imperfect presentation." Within these methodological boundaries, research in this area of scientific visualization has been carried out by a handful of researchers and [22] presents an updated comprehensive survey of the techniques introduced in the last decade to visualize uncertainty and errors in 3D.

True to the case, the GIS community carried out some of the earliest work on 3D representation of "visualization-induced" errors; in this field, indeed, the effects of uncertainty are of particular concern [99]. Woods and Fisher address some issues in 2D visualization of terrains and provide "gray-scaled maps" of the local root mean square error that are supposed to be paired with false color maps carrying the primary information of a scalar quantity. This quite naive approach sharply separates the data visualization from the uncertainty visualization and provides uncluttered and easy-to-appreciate maps. This technique moreover does not require any specialized software and can be carried out with the standard visualization tools available in any modern GIS. On the other hand, it simply disregards one of the fundamental driving motivations of modern scientific visualization: data integration. Treating uncertainty as an extra variable, increasing the dimension of the data set by one, is a simple legitimate solution but it is an unsatisfactory one; uncertainty is not a variable just like all the others in that it is associated with some measure of another variable. In order to provide to the observer some meaningful clues, the geometry of the variable under examination has to be integrated with the uncertainty information [51].

The need for a holistic approach to represent data, together with their uncertainty, has been also called "verity representation" by some researchers [85].

A taxonomy of the visualization techniques for uncertainty splits them into two great categories: intrinsic and extrinsic method. More precisely, intrinsic methodologies generally rely on augmenting the usual visual variables adopted in scientific visualization (position, size, brightness, texture, color, orientation, and shape) with variables such as boundary (of variable thickness, texture, and color), blur, transparency, animation, and extra dimensionality. Intrinsic techniques are more often associated with the representation of continuous (or to say it better "dense") data sets. Extrinsic methodologies include objects such as dials, thermometers, arrows, bars, objects of different shapes, and complex objects (glyphs). These last techniques are often associated with discrete (or to say it better "sparsely sampled") data sets. These approaches moreover pose a complex human perception problem: While a glyph may be appropriate by itself, the user's perception may be different when a group of glyphs is presented in various scales and locations [85]. In both intrinsic and extrinsic cases, some visual metaphors reveal themselves more appropriate than others to carry uncertainty information: dashed lines instead of solid ones, thick blurred lines instead of thin sharp ones, arrows attached to points and lines, blurred icons, and multiple over-imposed icons. The intrinsic-extrinsic paradigm by itself, however, does not provide a complete view of the field. A classification of visualization technique into dynamic and still schemes has been also proposed [76], and multimodal approaches like sonification [65] or animation [12] have to be taken into consideration to get a complete panorama of these research areas.

It is beyond the scope of this chapter to provide an exhaustive survey; instead, in the following subsections, the taxonomies mentioned above provide a general guide to review some published proposals for visualizing uncertainties that are more relevant to the GIS case. In particular, techniques to visualize uncertainty in flow computation and visualization that constitute a lively research subarea in this topic are intentionally left out for sake of brevity and to keep the focus on GIS.

6.2 INTRINSIC TECHNIQUES

In 1995, Wittenbrink [98] observed that interpolation is often at the core of data representation and analysis. Traditional interpolation schemes, unfortunately, may introduce an artificial "smoothness" in the data, conveying the wrong impression that data behave (at least locally) as well-behaved mathematical functions. To circumvent this, Wittenbrick proposed to use iterated fractal system (IFS) to interpolate between data. The resulting surfaces present a user-controlled degree of irregularity, but are, at the same time, constrained to the original data. The recursive nature of IFS moreover allows for fast rendering or for the adoption of IFS interpolation in substitution of traditional interpolation techniques locally and only in the subareas where it is needed. Wittenbrink shows that the roughness of the resulting surface is controlled with the choice of a few parameters that could be, in turn, selected according to the degree of uncertainty of the data to interpolate.

In 2003, Miller et al. [76] proposed a technique to visualize on a 2D map uncertainty linked with the results provided by a simplified mathematical model of the global water balance. A false color map is adopted to visualize water balance that is, of course, the most important variable of this study. For each pixel in the map, a hue is selected using a fixed color map, according to the value associated to that pixel. For example, a "cold-to-warm" color palette may be adopted to represent, respectively, low and high values of the scalar field that one wishes to map. Typically, a legend of the adopted color map next to the map makes the values immediately appreciable to the visual observer. To visualize the uncertainty affecting the estimate of the scalar field at each pixel, a variation of the luminance of the affected point is adopted, while its hue is fixed. The space of false colors that this technique eventually adopts may be represented by a triangle; the hue variation (the color map) is represented along one of its edges, and the variation in luminance is represented along the segments connecting this edge with the opposite vertex. This last vertex has a neutral hue and zero luminance value; it represents the maximum amount of uncertainty in the model. Since the technique is intrinsically 2D, room is left to introduce extrinsic objects (vertical bars) to enforce the uncertainty perception or to visualize other variables of the model.

6.3   EXTRINSIC TECHNIQUES

In 2000, Cedilnik and Rheingans [17] proposed an innovative approach called "procedural annotation." The basic idea is to augment the data visualization, adding to the "visual channels" ordinarily used to represent the data values a new "visual layer" whose perceptive characteristics are controlled in a procedural way according to the local amount of uncertainty. To be effective, this new layer (the "annotation") must obey some rules that Cedilnik summarizes as: "Procedurally generated — can be evaluated at any point of the image independently from other points. " Perceptually normalized — same amount of energy present at every place. Inherently meaningful — have some intuitive way of presenting the uncertainty." Reasonably fast – the generation of procedural annotations should not slow down the visualization. A concrete example of procedural annotation proposed by Cedilnik in the same paper is the following. Plane scalar data are represented in a rectangular region with a chosen color map. A rectangular grid (whose size is adjusted in order not to get perceptively over-important) of brighter lines is overimposed on this visualization. Cedilnik discusses several proposals: Lines may be thinned out, emboldened, perturbed by geometrical deformation, perturbed by punctual noise, and so forth. To get an idea of the good results that this approach may provide in the GIS area, a typical annotated map is reported in Figure 6.2, from Cedilnik [17].

In 2002, Olston and Mackinlay [84] discussed visualization issues related with uncertainty. Following the claim that generally uncertainty may be classified into two broad categories, statistical and bounded, they recommend the adoption of different visualization devices for the two cases. Statistical uncertainty arises when data are analyzed and summarized with some statistical procedure; in this case, what one wishes to represent is a more or less peaked density distribution. Bounded uncertainty, on the other hand, simply reports the range enclosing the observed data and avoids to make any other assumption on their distribution. Their analysis is exemplified in a very simple case: error-bars and box-plot. In this case, they recommend to use

 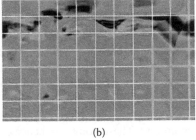

(a)   (b)

FIGURE 6.2 (a) Original data; (b) the grid annotation has been overimposed. The grid is thinned in less uncertain areas and thickened in the most uncertain ones (From Cedilnik and Rheingans, 2000. With permission).

an "ambiguated" form of these traditional glyphs to represent bounded uncertainty of univariate functions. This paper provides no real suggestion to generalize these proposals to higher dimensions; a rather pessimistic claim about the difficulty of solving this issue with extrinsic methods is its main conclusion.

In 2005, Botchen et al. [11] proposed a generalization of the traditional "spot noise" technique introduced in 1991 by van Wijk to represent uncertainty in flow visualization. The technique applies a procedural texture to the vector field. The content of high frequencies of the texture is inversely proportional to the uncertainty. The texture is hence blurred in areas of strong uncertainty and finely detailed in areas of small uncertainty. The basic texture is hence blurred and distorted according to the vector field that has been computed to describe the flow. The good results produced in this way, unfortunately, do not appear immediately useful in the GIS area, where dynamic representation of flows is not a common task.

6.4 MIXED TECHNIQUES

The uncertainty problem in interpolating a surface from a cloud of 3D points is an especially relevant case that has got major attention from the computer graphics community since the advent of fast and reliable 3D scanners. These devices provide unstructured (frequently noisy) point clouds that have to be cleaned and processed into a geometric surface representation. Several authors [9] have proposed to solve this problem, producing instead of a thin-sheet surface a "fat surface." This idea is certainly valid in reverse-engineering application, but to apply it to uncertainty visualization, an appreciation of the thickness of the fat surface has to be made possible. The only way to achieve it is through transparency. Transparency, in turn, may easily obfuscate a correct perception if the surface to model is geometrically complex.

To overcome this difficulty in 2002, Grigoryan and Rheingans [51] proposed a technique called "probabilistic surface." The mainstream computer graphics technique to visualize surfaces is to triangulate them and to send to the graphic hardware the description of the triangles coming from this transformation. Modern graphic processors are indeed optimized in dealing with triangles and, as such, they achieve amazing performances. As an alternative (or better, complementary to the "triangle primitive model"), simple primitives, other than triangles, have been proposed. The most popular of such graphic primitives is the "point." Images built mapping into the screen points have the typical "cloudy" appearance that makes them a natural choice for uncertainty representation. The probabilistic surface technique starts with a surface represented with a traditional triangular mesh. Assume that at any vertex of the mesh, an uncertainty measure has been provided. For each triangular facet, a random set of points lying on that facet is generated. The number of such points is user-controlled. These points are hence displaced, colored in a different manner from the supporting facet, or made transparent to some degree, according to the interpolated uncertainty value that can be assigned to each one of them. The user may choose to visualize this "cloud" of points together with the supporting triangulated surfaces or without it. The resulting images may be considered an "intrinsic" representation of the uncertainty, but since a point could be considered as a very elementary (yet powerful!) glyph, the method could be considered extrinsic as well. This technique has been applied to visualize boundaries of tumors in the human body, as a means to help surgeons in planning treatments, and in general, it is very effective to visualize 3D surfaces affected by uncertainty. To get complete advantage of the technique, however, the user has to be provided with some even limited possibility of navigating into the 3D space. The full 3D structure of the cloud is, indeed, best appreciated when the surface moves relative to the observer in a slow controlled way.

In conclusion, this section is perhaps a little too sketchy. As the above analysis shows there is a wide choice of methods and techniques that can be applied to the problem of uncertainty visualization. After about 10 years of research, it seems to be recognized by the mainstream that uncertainty cannot be put aside and that good scientific visualization has to include it, possibly in a holistic or, at least, well-integrated way with the other data. The above review should convince that there is no single method that is best suited for all GIS applications. When the interest is focused on the analysis of the data and this analysis does not require to go into 2D or into 3D, the annotation approach of [17] seems to be the most economic and elegant. When appreciation of volumes or of terrains is crucial, the top front methods seem to be all related to the proposal of [51] and to the idea of probabilistic surfaces. A general purpose GIS should perhaps include both approaches as a standard tool to help practitioners to integrate uncertainty representation into the illustrations of their data.

7 Applications

Cidália Fonte, Jorge Santos, Gil Gonçalves,
Salvatore Spinella, and Marcello Anile

CONTENTS

7.1   LINEAR INTERPOLATION — BACCHIGLIONE RIVER

Environmental regulations require the monitoring of the environmental state of a water basin in order to preserve and improve the water quality with respect to a target fixed in advance in terms of indexes. The necessity to represent a large amount of data related to an observation period and to compare it with respect to an ideal target (represented by a suitable index) compels to consider the uncertainty involved in the data survey and classification. The uncertainty arises from:

- lack of precision in environmental surveying,
- vagueness of the quality index definitions,
- arbitrariness in the classification with respect to the indices.

As pointed out by [89], thresholds for environmental indices are meaningful only in the context of the knowledge of natural background levels, regulatory policy, and the vulnerability of the main environmental components, and this implies that their values are comprised within some ranges according to some degree of confidence. Therefore, it is natural to relate environmental indices and their thresholds to some idea of acceptability measure, which would be restrictive to treat as a crisp number but instead could in a more natural way encompass some range. Hence, an environmental index or threshold could be interpreted as an interval or, more generally, as the membership in a fuzzy set that is constructed from a nested sequence of intervals together with a presumption level for each interval. The representation of uncertainty in an environmental survey by means of fuzzy number interpolation allows to obtain a continuous model of the pollutants along the main axis of a basin, which is useful to locate the pollution sources, to monitor them, and to operate in order to improve the environmental quality.

7.1.1 Fuzzy Representation of River Pollution

The proposed methodology was applied to a specific case study: the water quality monitoring of Bacchiglione River in the northeastern part of Italy. Water monitoring is carried out by Veneto Region Environmental Prevention and Protection Agency (ARPAV), the institutional body with the responsibility of environmental monitoring and control of pollution sources. The monitoring is referred to the period 2000–2004. Since January 1, 2000, the monitoring network has been reorganized according to the legal disposition of the Italian Decree no. 152/1999 so that the period that the data set analyzed can be considered homogeneous (the sampling frequency is monthly, the stations are not moved, and so forth.).

This work considers the Bacchiglione River as axis of the water basin. Along the river, there are seven monitoring places where monthly samples are collected and analyzed for tracing polluting substances like nitrogens, toxic metal, hydrocarbons, and bacteria. Therefore, river monitoring consists of a sampling sequence during 1 year, and the water quality in a basin is established with respect to a long period of observations. In this work, fuzzy representation was used in order to summarize the sampling observations and to represent the uncertainty of monitoring. The fuzzy number $\tilde{Z}$ collects 4 years of observations of a monitor point and has been constructed preserving the convex hull property as a map (α-level, interval), which associates to each α-level the smallest, median interval that contains a fraction $(1 - \alpha)$ of the set of data. Although a little simplistic, we have found that such a construction is sufficiently robust and compares well with statistical considerations. Then, these fuzzy numbers representing the sampled quantity were interpolated using cubic splines, which envelop the data in order to have a continuous model as a function of the distance to the mouth of the river (see [4]). The splines are finally interrogated at each kilometer using quality threshold. Figure 7.1 shows a representation of the interrogation involved in water monitoring. Notice that the resulting lines do not represent means

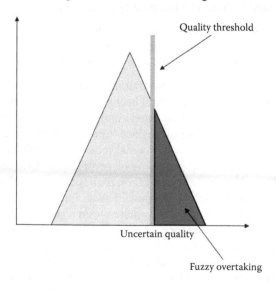

FIGURE 7.1 Fuzzy overtaking.

FIGURE 7.2 Bacchiglione River basin.

but the requirement that the quality threshold corresponding to each index is exceeded in the sense of fuzzy overtaking with $\delta = \frac{2}{3}$ (see [4] and Figure 7.2).

7.1.2 RESULTS

From Figure 7.3, a higher level of contamination in the lower part of the basin, due to urban sewage and agrochemical loads along the whole hydrographic basin, is evident. Figure 7.4 shows the two main contributions of civil and industrial loads with the peak of chemical oxygen demand (COD) as seen in the towns of Vicenza and Padua. The civil load is confirmed in Figure 7.5 with *Escherichia coli*, an indicator of fecal pollution (both of human and of animal origin). The river Bacchiglione presents a nitrate contamination in relation to agricultural leachate (Figure 7.6) in the middle and lower parts of the hydrographic basin. The biochemical oxygen demand (BOD5) (Figure 7.7) follows the same graph of Figure 7.4 for COD but in a more evident way: the peaks indicate contributions of the discharges first of the town of Vicenza and then of the town of Padua. Low values of oxygen saturation percentage, showed

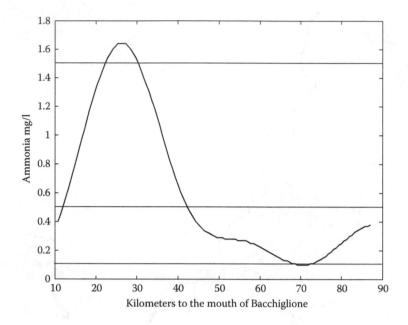

FIGURE 7.3 Ammonia.

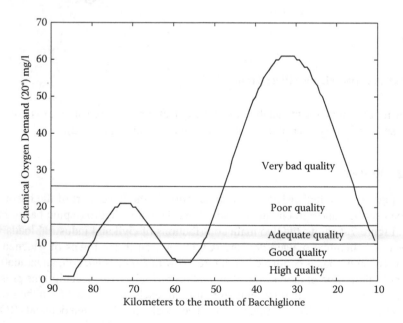

FIGURE 7.4 Chemical oxygen demand.

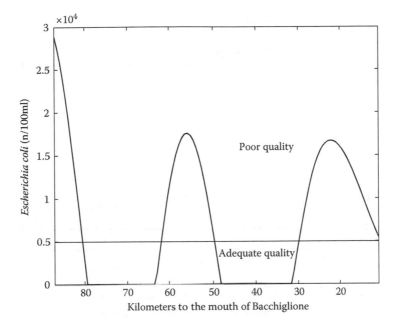

FIGURE 7.5 *Escherichia coli.*

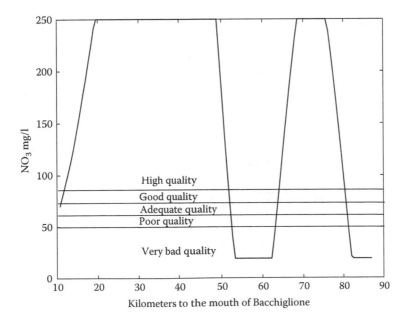

FIGURE 7.6 Nitrogens.

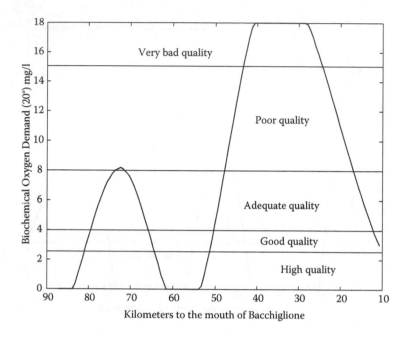

FIGURE 7.7 Biochemical oxygen demand.

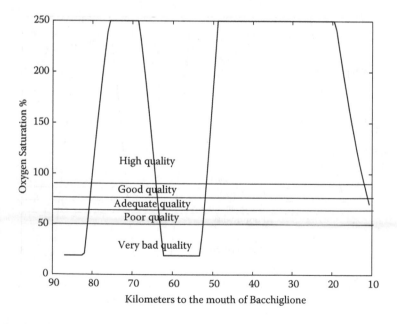

FIGURE 7.8 Oxygen saturation percentage.

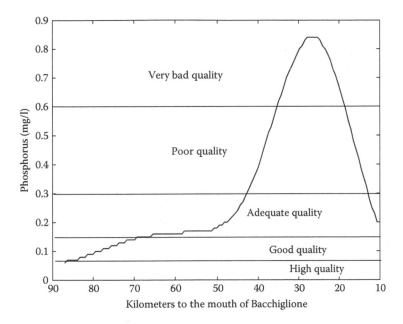

FIGURE 7.9 Phosphorus.

in Figure 7.8, is related to water pollution. High values could be related to water movements (falls, narrowing of the river, etc.). Figure 7.9 shows phosphorus values along the river that are related to the agricultural exploitation of the river basin lands. These human activities are more intense near the mouth of the river. The same graph is pointed out from Figures 7.10, 7.11, 7.12, and 7.13 in the period 2000–2004. Total coliforms (TC), fecal coliforms (FC), *Escherichia coli* (EC), and Enterococci are used as bacterial indicators for water quality monitoring and health assessment as each group of bacteria is prevalent in the intestines and feces of warm-blooded mammals, including wildlife, livestock, and human. These indicators are not pathogens: Fecal coli,

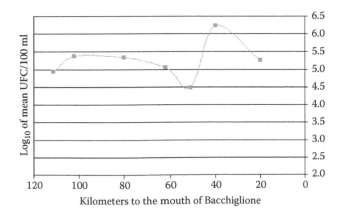

FIGURE 7.10 Total coliforms.

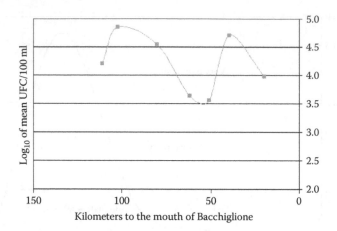

FIGURE 7.11 Fecal coliforms.

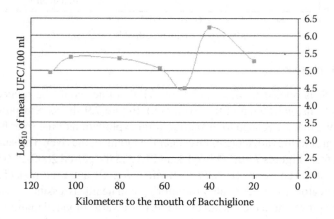

FIGURE 7.12 *Escherichia coli.*

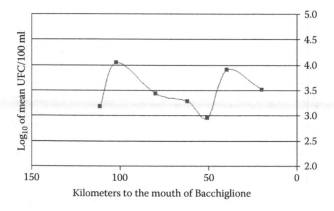

FIGURE 7.13 Fecal streptococci (FS).

fecal streptococci, and salmonella are used because they are less costly to detect and describe than the pathogens themselves.

In general, the Bacchiglione River basin is characterized with a high civil and industrial load and a significant agricultural load. The worst situation along the river is pointed out after the sewage system contributions for the towns of Vicenza (upper basin part) and Padua (lower basin part).

These results suggest that fuzzy interpolation is able to represent environmental state in terms of data measurements. Furthermore, fuzzy uncertainties can be queried to assess environmental quality from data directly, without statistical consideration about data.

7.2   AIR POLLUTION

Geographical data concerning environment pollution consist of a large set of temporal measurements (representing, e.g., hourly measurements for 1 year) at a few scattered spatial sites. In this case, the temporal data at a given site must be summarized in some form in order to employ them as input to build a spatial model. Summarizing the temporal data (data reduction) will necessarily introduce some form of uncertainty, which must be taken into account.

Statistical methods reduce the data to some moments of the distribution function as means and standard deviations, but these procedures rely on statistical assumptions on the distribution function, which are hard to verify in practice. In the general case, without any special assumption on the distribution function, statistical reduction can grossly misrepresent the data distribution. An alternative way is to represent the data with fuzzy numbers, which has the advantage of keeping the full data content (conservatism) and also of leading to computationally efficient approaches. This method has been employed for ocean floor geographical data by [40] (in the interval case) and [3] (for fuzzy numbers), and to environmental pollution data by [4].

Once the temporal data at the given sites have been summarized with fuzzy numbers, then it is possible to resort to fuzzy interpolation techniques in order to build a mathematically smooth deterministic surface model representing the spatial distribution of the quantity of interest. An alternative approach would be to employ fuzzy kriging, which will build a stochastic model. However, our aim is to construct a smooth deterministic model, because this could be used for simulation purposes.

We shall use fuzzy kriging only to estimate the missing information, which is required just outside the domain boundary, as we shall see, in order to build a consistent deterministic model.

7.2.1   FUZZIFICATION AND MAP CONSTRUCTION

The construction of a fuzzy map consists of the following steps:

Fuzzification of data: The fuzzy number $\tilde{Z}$ has been constructed preserving the *convex hull* property given in Equation (2.4) as a map (α-level, interval), which associates to each α-level the smallest *median interval* that contains a fraction $(1 - \alpha)$ of the set of data.

Kriging: Outside the region of measured data, "virtual" observations are estimated by kriging in order to complete the information needed for the approximation procedure.

Approximation: A smooth and deterministic model is fitted by a fuzzy B-spline.

For the kriging step, the following correlation function was chosen

$$C(h) = C_0 - e^{-\frac{h}{a}} \tag{7.1}$$

where C_0 is the variance of the fuzzy distribution. This is a simple assumption for an isotropic distribution of the data. Moreover, notice that the choice of an anisotropic correlation function for the kriging procedure can be used to take into account more complex characteristics of the region like orography and microclimate.

The data for CO referring to the city of Catania in the year 2001 have been represented by 17 fuzzy numbers through the α-level construction previously introduced. Each fuzzy number represents the measures of a sensor placed in the city. Eight kriged points were added at the vertexes and at the sides of the map, in order to make an estimation on the bound of the map. Then, we have chosen a regular grid

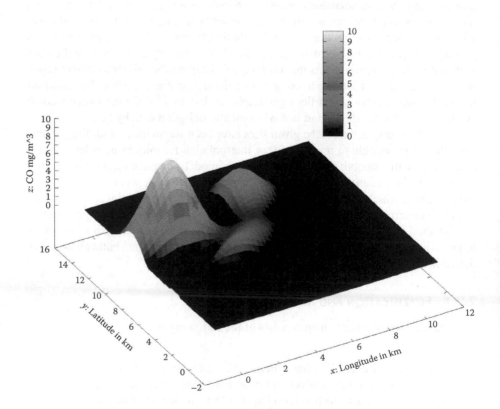

FIGURE 7.14 The z axis represents the level of pollution defuzzified by overtaking.

6×8 and constructed the fuzzy B-spline surface of order 4 by solving the problem in Equation (2.17). At last the fuzzy map was interrogated at growing level of CO.

Figure 7.14 shows on the z axis the level of CO pollution exceeded in the mapped urban area (in the sense of Definition 1.20).

In this case, the fuzzy kriging procedure has been used only to supply missing information just outside the boundary of the region of interest. However, one could envisage a hybrid method where the kriging procedure would create a regularly spaced set of fictitious data observations in the form of fuzzy numbers, which afterward would be approximated by a fuzzy B-spline, in order to construct a viable smooth deterministic model.

7.3 IDENTIFICATION OF ZONES SUITABLE FOR CONSTRUCTION USING FUZZY GEOGRAPHICAL ENTITIES

As explained in chapter 4, fuzzy geographical entities (FGEs) can be constructed using several approaches within several contexts. Different types of information can be used, and therefore the resulting FGEs may have different semantics.

The present example illustrates an application where FGEs are built and a simple analysis is performed with those entities. The aim is to identify the zones suitable for constructing an infrastructure. The regions must have relatively small slopes and good solar exposure. In a traditional crisp analysis, the conditions imposed could be slopes between 0 and 10 percent and more or less faced to south, for example, with an aspect between 135 and 225 degrees (considering 0 degrees in the north direction).

A digital terrain model with 10-meter resolution was used. The slope and aspect of the region under study were computed and zones corresponding to the criteria described above identified. These regions can be seen, respectively, in Figures 7.15 and 7.16. To obtain the region satisfying both conditions, an overlay of the above results is done, determining the intersection of the obtained regions. The resulting regions are shown in Figure 7.17.

Notice that, with this crisp approach, regions with slopes a little larger than 10 percent are not considered as adequate at all, and the same occurs to regions with aspect smaller than 135 degrees or larger than 225 degrees. For example, a zone with slope equal to 11 percent and aspect of 226 degrees is not considered appropriate, while a zone with 10 percent slope and aspect equal to 225 degrees is considered as good as any other. This abrupt transition between the regions considered as convenient, and the ones excluded à priori from the analysis, can be avoided if, instead of considering the appropriate values of slope and aspect with crisp sets, fuzzy sets are used. Figures 7.18 and 7.19 show example membership functions that can be used to consider the slope and aspect values appropriate for the study.

Degrees of membership to the "convenient slopes for construction" and "regions with good solar exposure" are computed for all regions. The results are shown in Figures 7.20 and 7.21. The overlay of these results is done with the fuzzy logical operator for intersection, and the results are shown in Figure 7.22 where, for each cell, a degree of suitability for construction is obtained. Figure 7.23 shows all regions with grades of membership between 0 and 1, that is, all regions that have not been

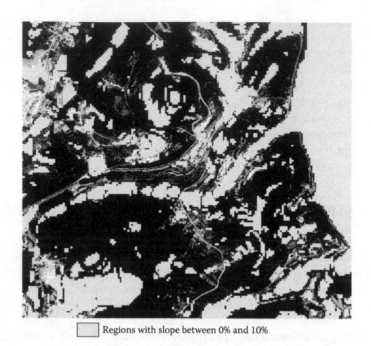

☐ Regions with slope between 0% and 10%

FIGURE 7.15 Regions with slope between 0 and 10 percent.

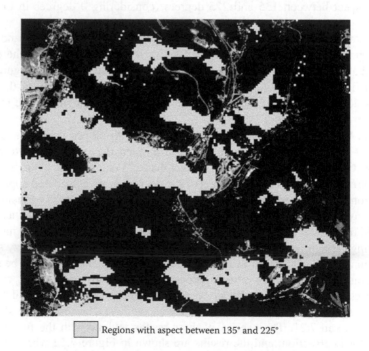

☐ Regions with aspect between 135° and 225°

FIGURE 7.16 Regions heading south (aspect between 135 and 225 degrees).

Regions with slope between 0% and 10% and aspect between 135° and 225°

FIGURE 7.17 Crisp zones suitable for construction (relatively small slopes and good solar exposure).

identified in the crisp analysis and were identified in the fuzzy one. Some of these regions have relatively good conditions, since they have grades of membership close to 1 (177 cells have grades of membership between 0.9 and 1, and 1159 between 0.7 and 1).

A similar analysis could be performed considering some crisp subintervals for slope values between 10 and 20 percent, and for aspect from 90 to 135 degrees and from 225 to 270 degrees. If, for example, three subintervals were considered inside each mentioned interval, after the overlay, nine new classes would be formed, resulting

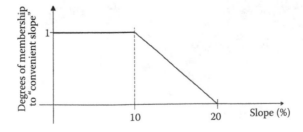

FIGURE 7.18 Membership function of the slope values to "convenient slopes for construction."

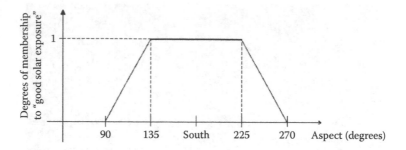

FIGURE 7.19 Membership function of the aspect values to "regions with good solar exposure."

from combinations between these classes. To take decisions based on this information, degrees of preference needed to be applied to each class, resulting in something similar to the result immediately obtained with the fuzzy analysis. Even though some similar analysis could be done using the crisp subintervals for the less suitable conditions, the main disadvantage of this crisp approach is that abrupt transition between the classes would always be present, which results in assigning different values of suitability to regions with similar characteristics because they are located close to the borders of

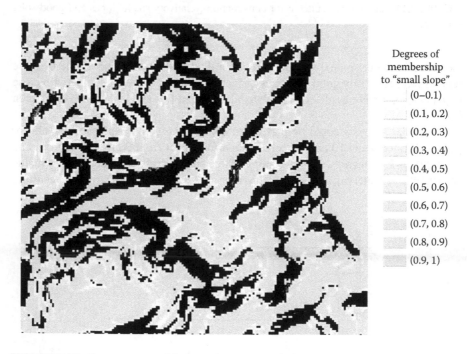

FIGURE 7.20 Fuzzy geographical entity corresponding to "convenient slopes for construction."

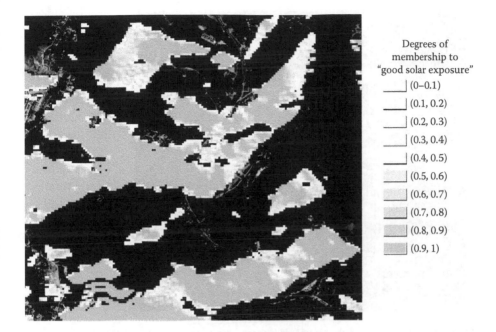

Degrees of
membership to
"good solar exposure"

	(0–0.1)
	(0.1, 0.2)
	(0.2, 0.3)
	(0.3, 0.4)
	(0.4, 0.5)
	(0.5, 0.6)
	(0.6, 0.7)
	(0.7, 0.8)
	(0.8, 0.9)
	(0.9, 1)

FIGURE 7.21 Fuzzy geographical entity corresponding to "regions with good solar exposure."

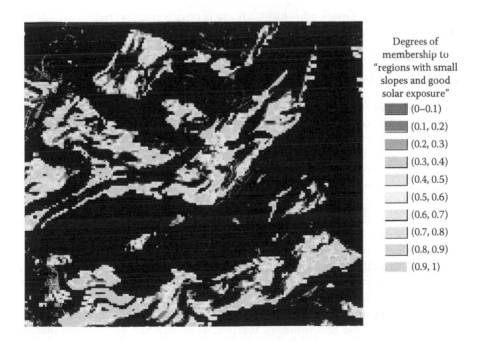

Degrees of
membership to
"regions with small
slopes and good
solar exposure"

	(0–0.1)
	(0.1, 0.2)
	(0.2, 0.3)
	(0.3, 0.4)
	(0.4, 0.5)
	(0.5, 0.6)
	(0.6, 0.7)
	(0.7, 0.8)
	(0.8, 0.9)
	(0.9, 1)

FIGURE 7.22 Fuzzy regions with relatively small slopes and good solar exposure.

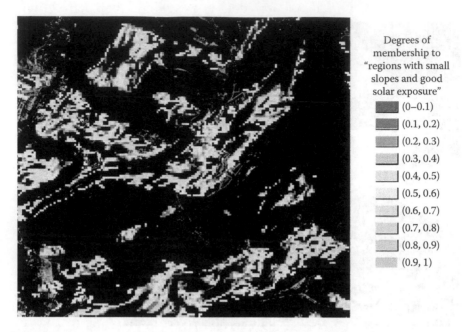

FIGURE 7.23 Regions with degrees of membership to "zones with relatively small slopes and good solar exposure" larger than 0 and smaller than 1.

contiguous intervals. On the other hand, a fuzzy approach enables gradual transitions to be considered and consequently a more realistic analysis.

7.4 UNCERTAINTY IN TERRAIN SLOPE CLASSIFICATION

The classification of topographic surface represented by a digital elevation model (DEM) into specific slope classes is a common task in GIS applications. However, the computation of these classes from a DEM is influenced by the uncertainty in the elevations of those models. Monte Carlo method has been used to study the effect of DEM uncertainty on topographic parameters. This method requires intensive computation, which turns the implementation very hard in most GIS software packages. Interval arithmetic (IA) has been used as a successful alternative to the Monte Carlo method. Intervals of variation are used instead of a significant number of simulations. The generalization from IA to fuzzy numbers is straightforward if we consider the interval as the fuzzy number support. The results from both methods show that the propagation of DEM uncertainty to the calculation of slope classes using IA is an interesting alternative to Monte Carlo method, since it does not require much intensive computation, which turns the implementation easier in GIS analyses [50].

Slope classification is affected by elevation uncertainty in DEM, which is generated by errors in the acquisition of topographical data and in the interpolation methods used to build the elevation model. The uncertainty includes errors or uncertainties due to imperfections of measurement systems and also the effect of the cartographic generalization, which cannot be avoided in cartographic modeling. In order that information about the quality of the topographic surface can be accessed by GIS users during spatial analysis, the following procedure has to be executed:

1. build an elevation uncertainty model;
2. propagate that uncertainty to derived terrain features (slope, aspect, curvature . . .);
3. specify appropriate methods for uncertainty evaluation, including visualization.

Here, we use a simple uncertainty model of cartographic terrain representation, which states that for a terrain representation based on contour lines and point elevations, 90 percent of control point elevations should have a mean-squared error not bigger than half of the contour interval. Here, we assume that DEM uncertainty can be modeled adding to every grid elevation a disturbing term from a random field spatially independent from neighbors. The Monte Carlo simulations are generated by the simple formula:

$$z(x, y) = z_0(x, y) + N(\mu_z, \sigma_z), \qquad (7.2)$$

where the mean μ_z is 0 and $\sigma_z = 0.304\Delta z$, following the previous uncertainty model (Δz is the contour interval). Interval elevations are considered here confidence intervals based on the same standard deviation σ_z.

Both methods were applied to the region in Figure 7.24. The results are condensed in Figure 7.25 and in Table 7.1.

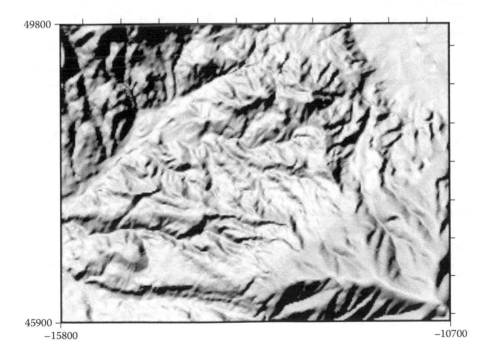

FIGURE 7.24 Shaded relief of study area.

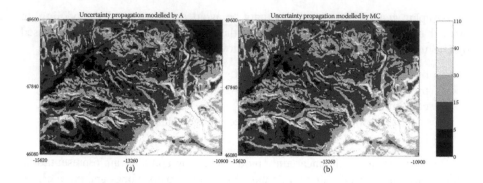

FIGURE 7.25 (a) Midpoint of interval slope; (b) mean value of Monte Carlo simulations.

TABLE 7.1
Confusion Matrix Comparing Uncertainty in Slope Classification Using Monte Carlo Method and Interval Arithmetic

		Slope classes with uncertainty — Monte Carlo method						
	Classes	I	II	III	IV	V	Total	PCC
	I	3755	3244	0	0	0	6989	53.73
Slope classes with	II	0	18266	496	0	0	18762	97.36
uncertainty — interval	III	0	0	12654	89	0	12743	99.3
arithmetic	IV	0	0	0	3770	43	3813	98.87
	V	0	0	0	0	3882	3882	100
	Total	3755	21500	13150	3859	3925	46189	91.64

The confusion matrix in Table 7.1 shows that the amount of cells classified in the same way by both methods is in fact 92 percent. Furthermore, if class I is ignored, the amount of pixels classified in the same way for each class is very high (more than 97 percent). This means that both methods of uncertainty propagation produce almost the same results for classes II, III, IV, and V. However, for class I, there is only 54 percent of agreement. This smaller agreement is due to the strong variation of elevation frequencies around the 5-percent value, which induces instability when class limit is close to that region of the histogram. For example, if we change classes I and II to 0–7 percent and 7–10 percent, respectively, we will get the values 82 and 96 percent of agreement, respectively. Therefore, the IA is an alternative to evaluate elevation uncertainty propagation to slope classification.

8 Algorithms — Pseudo Code

Cidália Fonte, Jorge Santos,
and Salvatore Spinella

CONTENTS

8.1 PSEUDO CODE FOR FUZZY SURFACES

8.1.1 CONSTRUCTION OF A FUZZY B-SPLINE

The construction of a fuzzy B-spline can be done with a resolution of a linear programming problem. In fact, the pseudo code 1 shows a single statement to perform the resolution of the linear programming problem in Equation (2.17). It must be noticed that this problem can be hard to solve numerically because it could involve thousands of variables and constraints, but this is not unusual in linear programming. However, many traps could fail the computation. Basically correlations among geographical data (which are very frequent) can lead to linear combination and could block the start of many linear routines. An easy solution to these problems can be a simple exchange of the rows into the matrix of the linear problem. This easy heuristic can help the pivoting strategies.

Algorithm 1   B-spline construction

Require: Observation $((x_i, y_i), F_i)$, xGridSpace, yGridSpace
  solve minimization problem in Equation (2.17) to envelop
  observations return solution as fuzzy numbers for grid knot

---

Another important point is the grid spacing of the B-spline. It must be noticed that fine grids introduce nodes that relax the surface on the constraints that is, reduce the uncertainty propagation.

The fuzzy B-spline evaluation on a point (x, y) is a simple summation of the products between $F(i, j)$ and the respective base $base_{i,j}$ computed in (x, y). The pseudo code 2 expounds this schema.

8.1.2   FUZZY OVERTAKE PSEUDO CODE

The concept of overtake is easy to implement because it requires only direct computation. The pseudo code 3 shows the summation with respect to a weight function g.

Algorithm 2   B-spline evaluation on a $M \times N$

Require:(x, y) to evaluate
  $acc \leftarrow 0$
  for $i \leftarrow 1$ to N **do**
    for $j \leftarrow 1$ to M **do**
      $acc \leftarrow acc + base(i, j, x, y)F(i, j)$
    end for
  end for
  return acc

Algorithm 3   Fuzzy overtake o between G and F

---

  for all α-level **do**
    compute the overtake o_α between intervals G_α and F_α using Equation (1.81)
  end for
  $o \leftarrow \int_\alpha o_\alpha * g(\alpha)$

Algorithm 4   B-spline interrogation: $t > F(x, y)$

Require: a threshold t (real or fuzzy), a fuzzy surface $F(x, y)$, a $\delta \in [0, 1]$
  if $Overtake(t, F(x, y)) > \delta$ **then**
    return TRUE
  else
    return FALSE
  end if

The algorithm for interrogation in section 8.1.2 is a predicate that returns "true" if there is an overtake of δ between fuzzy numbers. A global interrogation on a map can be done with one of the many algorithms of *scan line*.

8.2   PSEUDO CODE FOR FUZZY INTERPOLATORS

8.2.1   CONSTRUCTION OF FUZZY UNIVARIATE LINEAR SPLINES

The construction of a fuzzy univariate linear spline is based on crisp linear splines. Every α-level limit is a crisp linear spline (see routine *flspline* in $fuzinterp$ MATLAB toolbox).

Algorithm 5   Fuzzy univariate linear spline

Require: $(xd, \tilde{z}d)$ and evaluation positions x
   for $i \leftarrow 1$ to *NumOfPositions* **do**
      for $\alpha \leftarrow 0$ to 1 **do**
         $FuzLinSp_i(\alpha) \leftarrow [\ LinearSpline(xd, z^-(\alpha); x_i),\ LinearSpline$
         $(xd, z^+(\alpha); x_i)]$
      end for
   end for
   return $FuzLinSp(x)$

8.2.2   CONSTRUCTION OF FUZZY UNIVARIATE CUBIC SPLINES

The construction of a fuzzy univariate cubic spline is based on basis cubic splines. Since these functions can be negative, consistency is to be achieved by solving a semi-infinite optimization problem. However, for computational reasons, it is much easier to find an approximation. Therefore, we suggest to use crisp cubic splines to approximate the α-levels' limits of the inconsistent fuzzy cubic splines (see routine *cfuzpline* in *fuzinterp* MATLAB toolbox).

Algorithm 6   Fuzzy univariate cubic spline

Require: $(xd, \tilde{z}d)$ and evaluation positions x
   for $i \leftarrow 1$ to *NumOfPositions* **do**
      for $\alpha \leftarrow 0$ to 1 **do**
         $FuzCubSp_i(\alpha) \leftarrow |\ BasisCubicSplines(xd; x_i)| * \tilde{z}d(\alpha)$
      end for
   end for
   for $\alpha \leftarrow 0$ to 1
      $ConsistFuzCubSp(\alpha) \leftarrow [\ FittedCubSp(x, FuzCubSp^-(\alpha)),$
      $FittedCubSp(x, FuzCubSp^+(\alpha))]$
   end for

for $i \leftarrow 1$ to $NumOfPositions$ **do**
  for $\alpha \leftarrow 0$ to 1 **do**
    $ConsistFuzCubSp_i(\alpha) \leftarrow ConsistFuzCubSp(\alpha; x_i)$
  end for
end for
return $Consist FuzCubSp(x)$

8.2.3 CONSTRUCTION OF FUZZY TINS

The construction of a fuzzy TIN just needs to triangulate the data points and find the planes that are the α-levels' limits for every triangle (see routine *ftin* in *fuzinterp* MATLAB toolbox).

Algorithm 7   Fuzzy TIN

Require: $(xd, yd, \tilde{z}d)$ and evaluation positions (x, y)
  $tri \leftarrow$ triangulation(xd, yd)
  for $i \leftarrow 1$ to $NumOfPositions$ **do**
    $T \leftarrow$ Find Triangle in tri containing (x_i, y_i)
    for $\alpha \leftarrow 0$ to 1 **do**
      $plane_T(\alpha) \leftarrow [plane(x_T, y_T, z_T^-(\alpha)), plane(x_T, y_T, z_T^+(\alpha))]$
      $FuzTIN_i(\alpha) \leftarrow plane_T(\alpha, x_i, y_i)$
    end for
  end for
  return $FuzTIN(x, y)$

8.2.4 CONSTRUCTION OF FUZZY SHEPARD METHOD

The construction of fuzzy Shepard method needs to find a neighborhood of interpolation positions to data points and the distances between them, to calculate the weights (see routine *fshep* in *fuzinterp* MATLAB toolbox).

Algorithm 8   Fuzzy Shepard method

Require: $(xd, yd, \tilde{z}d)$ and evaluation positions (x, y)
  for $i \leftarrow 1$ to $NumOfPositions$ **do**
    $neighs \leftarrow$ neighbors(xd, yd) of (x_i, y_i)
    $weights \leftarrow$ distances of (x_i, y_i) to $neighs$
    for $\alpha \leftarrow 0$ to 1 **do**
      $FuzShep_i(\alpha) \leftarrow weights * \tilde{z}(\alpha, neighs)$
    end for
  end for
  return $FuzShep(x, y)$

8.2.5 CONSTRUCTION OF FUZZY THIN-PLATE SPLINES

The construction of fuzzy thin-plate splines also needs to find a neighborhood of inter-polation positions to data points. The weights here are determined from the solution of a linear system of radial functions. Since weights can be negative here, consistency is to be achieved also by solving a semi-infinite optimization problem. As in Algorithm 6, it is much easier and more efficient to find an approximation. Therefore, we suggest to use crisp thin-plate splines to approximate the α-levels' limits of the incon-sistent fuzzy thin-plate splines (see routine *cftps* in *fuzinterp* MATLAB toolbox).

Algorithm 9   Fuzzy thin-plate splines

Require: $(xd, yd, \tilde{z}d)$ and evaluation positions (x, y)
  for $i \leftarrow 1$ to *NumOfPositions* **do**
    neighs $\leftarrow$ neighbors(xd, yd) of (x_i, y_i)
    weights $\leftarrow$ SolveLin$((x_i, y_i), neighs)$
    for $\alpha \leftarrow 0$ to 1 **do**
      $FuzTPS_i(\alpha) \leftarrow weights * \tilde{z}(\alpha, neighs)$
    end for
  end for
  return $FuzTPS(x, y)$

8.2.6 CONSTRUCTION OF FUZZY CUBIC SPLINES

The construction of a fuzzy cubic spline surface is based on tensor product of basis cubic splines in x and y. Again, the consistency is to be achieved by solving a semi-infinite optimization problem, but the best approach is also to find an approximation using crisp cubic splines to approximate the α-levels' limits of the inconsistent fuzzy cubic splines (see routine *cfuzpline2* in *fuzinterp* MATLAB toolbox).

Algorithm 10   Fuzzy cubic spline

Require: $(xd, yd, \tilde{z}d)$ and evaluation positions (x, y)
  for $i \leftarrow 1$ to *NumOfPositions* **do**
    for $\alpha \leftarrow 0$ to 1 **do**
      $FuzCubSp_i(\alpha) \leftarrow | BasisCubicSplines(xd, yd; x_i, y_i)| * \tilde{z}d(\alpha)$
    end for
  end for
  for $\alpha \leftarrow 0$ to 1 **do**
    $ConsistFuzCubSp(\alpha) \leftarrow [FitCubSp(x, y, FuzCubSp^-(\alpha)),$
    $FitCubSp(x, y, FuzCubSp^+(\alpha))]$
  end for

```
for i ← 1 to NumOfPositions do
  for α ← 0 to 1 do
    ConsistFuzCubSp_i(α) ← ConsistFuzCubSp(α; x_i, y_i)
  end for
end for
return ConsistFuzCubSp(x, y)
```

8.3 PSEUDO CODE FOR THE CONSTRUCTION OF FUZZY GEOGRAPHICAL ENTITIES

As explained in chapter 4, several approaches can be used to construct FGEs, depending on what they represent and the information available. Algorithms to build FGEs over a tessellation are presented below, considering the four possible sources of uncertainty mentioned in chapter 4. For all cases, a tessellation of the geographical space is considered, and the value z of the characteristic used to define the attribute is known in each elementary region of the tessellation.

8.3.1 UNCERTAINTY IN THE ATTRIBUTE DEFINITION

This algorithm can be used to construct FGEs when there is uncertainty in the definition of the attribute characterizing the GEs.

Algorithm 11   Uncertain attribute

Require: z_j values for all elementary regions r_j of the tessellation and membership function $\mu_A(z)$ characterizing attribute A

```
for all z_j values do
  μ_{FGE_A}(r_j) ← μ_A(z_j)
end for
```

8.3.2 SEVERAL VERSIONS OF THE ATTRIBUTE

This algorithm applies when there are several versions of the attribute characterizing a GE, resulting in uncertainty in the attribute definition.

Algorithm 12   Several versions of the attribute

Require: z_j values for all elementary regions r_j of the tessellation, i sets $(Z_A)_i$ corresponding to all versions (Z_A and the degree of confidence of each version i of A)

```
for each elementary region r_j do
  counter ← 0
  for each version i do
    if z_j belongs to (Z_A)_i then
```

$\qquad counter \leftarrow counter + \text{degree of confidence of version } i$
    end if
  end for
  $\mu_A(z_j) \leftarrow counter$
  $\mu_{FGE_A}(r_j) \leftarrow \mu_A(z_j)$
end for

The degrees of confidence assigned to each version of the attribute must be a value between 0 and 1, and the sum of the degrees of confidence assigned to all versions of the same attribute must add up to 1.

8.3.3 ERRORS IN THE BASE ATTRIBUTE VALUES

This algorithm applies when the values of the base attribute at the elementary regions are affected by errors. These errors can be used to construct FGEs.

Algorithm 13   Uncertainty in z values

Require: z_j values for all elementary regions r_j of the tessellation, the attribute values $Z_A = [\underline{z}, \overline{z}]$ defining attribute A, the error function of each z_j: $EF(z_j)$ and the maximum and minimum values z_j can take z_{max} and z_{min}

$$\mu_{FGE_A}(r_j) \leftarrow \frac{\int_{\underline{z}}^{\overline{z}} EF_{(z_j)} dz_j}{\int_{z_{min}}^{z_{max}} EF_{(z_j)}\, dz_j}$$

8.3.4 EVOLUTION OVER TIME

This algorithm applies when the geographical locations of the GEs change over time. The information about the position of the entities in different epochs can be used to construct FGEs.

Algorithm 14   Evolution over time

Require: all versions of the geographical entity characterized by A and the degree of confidence of each version
  for each elementary region r_j **do**
    $counter \leftarrow 0$
    for each version **do**
      if region belongs to version **then**
        $counter \leftarrow counter + \text{degree of confidence of version}$
      end if
    end for
    $\mu_{FGE_A}(r_j) \leftarrow counter$
  end for

8.4 PSEUDO CODE FOR OPERATIONS WITH FUZZY GEOGRAPHICAL ENTITIES

8.4.1 Fuzzy Area Computation

The fuzzy area determination requires the computation of the crisp area of several α-levels of the FGEs. The α-levels may be chosen by the user according to the needs and the variation of the FGEs grades of membership. In Matlab routine fuzzyarea.m found in the attached CD, default values α are 0.001, 0.1, 0.2, 0.3, 0.4, 0.5, 0.6, 0.7, 0.8, 0.9 and 1. The pseudo code algorithm 15 shows the basic steps of the fuzzy area routine.

Algorithm 15 Fuzzy area computation

Require FGE E, set of values $\alpha \in [0, 1]$, cell size
   for all α-levels **do**
      $A(\alpha) \leftarrow$ area of α-level $^\alpha E$
   end for
   for $i \leftarrow 1$ to number of α values **do**
      $slope(i) \leftarrow$ slope of the segment defined by points $[A\{\alpha(i)\}, \alpha(i)]$ and
      $[A\{\alpha(i+1)\}, \alpha(i+1)]$ $y-$ intersect $(i) \leftarrow y-$ intersect of the segment defined
      by points $[A\{\alpha(i)\}, \alpha(i)]$ and $[A\{\alpha(i+1)\}, \alpha(i+1)]$
   end for
   eliminate unnecessary branches of membership function
   return *slope, y-intersect* and points (A, α) defining each branch of the fuzzy area

---

8.4.2 Fuzzy Perimeter Computation

Pseudo code algorithm 16 shows the basic steps of the computation of the fuzzy perimeter of FGEs. Such as with the fuzzy area algorithm, the fuzzy perimeter determination also requires the computation of the crisp perimeter of several α-levels of the FGE. In Matlab routine fuzzyperimter.m found in the attached CD, default values of α are 0.001, 0.1, 0.2, 0.3, 0.4, 0.5, 0.6, 0.7, 0.8, 0.9 and 1.

Algorithm 16 Fuzzy perimeter computation

Require FGE E, set of values $\alpha \in [0, 1]$, cell size
   for all α-levels **do**
      $P(\alpha) \leftarrow$ perimeter of α-level $^\alpha E$
   end for
   for $i \leftarrow 1$ to number of α values-1 **do**
      $slope(i) \leftarrow$ slope of the segment defined by points $[P\{\alpha(i)\}, \alpha(i)]$ and
      $[P\{\alpha(i+1)\}, \alpha(i+1)]$ $y-$ intersect $(i) \leftarrow$ y-intersect of the segment defined
      by points $[P\{\alpha(i)\}, \alpha(i)]$ and $[P\{\alpha(i+1)\}, \alpha(i+1)]$
   end for
   fuzzyperimeter$(P) = $ max (grade of membership considering all branches defined
   by points $[P\{\alpha(i)\}, \alpha(i)]$ and $[P\{\alpha(i+1)\}, \alpha(i+1)]$ to which P may belong)
   return *slope, y-intersect* and points (P, α) defining each branch of the fuzzy
   perimeter

---

9 Appendices — Fuzzy Arithmetic and Fuzzy Query C++ Source

Salvatore Spinella

CONTENTS

DISCLAIMER OF WARRANTY

We make no warranties, expressed or implied, that the programs contained in this volume are free of error, or are consistent with any particular standard of merchantability, or that they will meet your requirements for any particular application. They should not be relied on for solving a problem whose incorrect solution could result in injury to a person or loss of property. If you do use the programs in such a manner, it is at your own risk. The authors and publisher disclaim all liability for direct or consequential damages resulting from your use of the programs.

9.1   A FUZZY NUMBER IMPLEMENTATION IN C++

This section introduces a base of C++ code for fuzzy number implementation. The implementation is based on a class named *FuzzyNumber*, which includes constructors and arithmetic operators. The member of this class is a map that implements the functional relation between α-level and interval. Many other operators (like inserter and extractor for input–output operations) and fuzzy function are listed in the code. Notice that all code is written as template with respect to a T type, which can assume integer type or floating point type.

```
/*
    Name: Fuzzy.h
    Description: A collection of templates to define in
    C++ the fuzzy arithmetic.
*/

#include "Interval.h"
#include <utility>
#include <vector>
```

```cpp
#include <map>
#include <stack>

/*
  Definition of Fuzzy Number
*/
 template<class T> class FuzzyNumber{ public:
     typedef T valuetype;

     FuzzyNumber(vector<T>&, vector<T>&);
     // Constructors
     FuzzyNumber();
     FuzzyNumber(T&);

     bool Insert(T&, Interval<T>&); // Insert of an
interval of level alpha

     FuzzyNumber<T>& operator+=(const FuzzyNumber<T>&);
     // Operators
     FuzzyNumber<T>& operator-=(const FuzzyNumber<T>&);
     FuzzyNumber<T>& operator*=(const FuzzyNumber<T>&);
     FuzzyNumber<T>& operator/=(const FuzzyNumber<T>&);

     pair<T,Interval<T> > operator()(const T&) const;
     // Map operator

     FuzzyNumber<T>& operator=(const T& a);
     // Real assignment
     FuzzyNumber<T>& operator+=(const T& a);
     // unary operator
     FuzzyNumber<T>& operator-=(const T& a);
     FuzzyNumber<T>& operator*=(const T& a);
     FuzzyNumber<T>& operator/=(const T& a);

     map<T,Interval<T> > A; // Map representation of a
};                              // Fuzzy Number


/*
Constructor of the fuzzy number zero at alpha-level zero
*/
template<class T> FuzzyNumber<T>::FuzzyNumber(){
     T zero=T();
     A[zero]=zero;
};
```

```
/*
 Constructor from real data of a fuzzy number given a
 vector of data and a vector of levels.
It constructs an alpha-level representation of the data.
Data are ordered  and for each  alpha level a median
interval, which envelops a portion (1-alpha) of data.
This choice supposes data are representative of the
uncertainty.

*/
template<class T>FuzzyNumber<T>::FuzzyNumber(vector<T>&
data, vector<T> & level)
    {
     T zero=T(), uno=T(1), l;
     unsigned n=data.size(),k;

     sort(data.begin(),data.end(),less<T>());

     A[zero] = Interval<T>(data[0], data[n-1]);

     for (unsigned i=0; i<level.size(); i++)
        {
          if (zero<level[i] && level[i]<uno)
            {
             k=(unsigned int)(level[i]*n/2);
             l=(T)(2*k)/n;
             A[l]=Interval<T>(data[k], data[n-k-1]);
            }
        }
    }

/*
 alpha-level inserting
*/
template<class T> bool FuzzyNumber<T>::Insert(T& level,
Interval<T>& i){

  typedef map<T,Interval<T> >::iterator CI;
  pair<CI, bool> bins=A.insert(pair<T, Interval<T> >
  (level, i));
  CI iter;

  if(!bins.second)return false;

  if((iter=bins.first) != A.begin()){
                    iter--;
```

```
                              if (!bins.first->second.in(iter->
                              second)) {
                                  A.erase(bins.first);
                                  return false;
                              }
      }

      iter=bins.first; iter++;
      if (iter!=A.end() && !iter->second.in(bins.first->
      second)){
                                  A.erase(bins.first);
                                  return false;
                              }
         return true;
}
```

```
/*
 Definition of operation "+=" between fuzzy numbers.
 This operator evaluates the representation of two fuzzy
 numbers and updates a result that represents the sum
 of the fuzzy numbers.
*/
template<class T> inline FuzzyNumber<T>& FuzzyNumber<T>
:: operator+=(constFuzzyNumber<T>& a)
   {
   typedef map<T,Interval<T> >::const_iterator CI;

   CI f1_end = a.A.end(), f1 = a.A.begin(),
      f_end = A.end(), f = A.begin();

    Interval<T> i=f->second, i1=f1->second;
    T alpha=f->first, alpha1=f1->first;

     for(;;){
      if(f->first > f1->first){
                 if(f1!=f1_end){
                                  A[alpha1] = i+i1;
                 if(++f1!=f1_end) {alpha1=f1->first;
                 i1=f1->second;}
                 else{ f++; A.erase(alpha);
                     if(f==f_end)break;
                     else{i=f->second; alpha=f->first;}
                 }
            }else {
```

```
                              A[alpha] = i+i1;
                              if(f!=f_end && ++f!=f_end){i=f->
                              second; alpha=f->first; }
              else break;
                          }
      } else if(f->first < f1->first){
                          if(f!=f_end){
                                      A[alpha] = i+i1;
                                      if(++f!=f_end){i=f->
second; alpha=f->first;}
                          else{ f1++; A.erase(alpha1);
                            if(f1==f1_end)break;
                            else {alpha1=f1->first; i1=f1->
                            second;}
                      }
                          }else {
                              A[alpha1] = i+i1;
                              if(f1!=f1_end && ++f1!
=f1_end) {alpha1=f1->first; i1=f1->second;}
                  else break;
                              }
      }else {
              A[alpha] = i + i1;
              if(f1!=f1_end && ++f1!=f1_end){alpha1=f1->
              first; i1=f1->second;}
              if(f!=f_end && ++f!=f_end){ alpha=f->first;
              i=f->second; }
              if(f1==f1_end && f==f_end)break;
              }
      }
      return * this;
}


/*
 Definition of operation "-=" between fuzzy numbers.
This operator evaluates the representation of two fuzzy
numbers and updates a result that represents the sum of
the fuzzy numbers.
*/
template<class T> inline FuzzyNumber<T>& FuzzyNumber<T>
::operator-=(constFuzzyNumber<T>& a)
   {
   typedef map<T,Interval<T> >::const_iterator CI;

   CI f1_end = a.A.end(), f1 = a.A.begin(),
      f_end = A.end(), f = A.begin();
```

```
    Interval<T> i=f->second, i1=f1->second;
    T alpha=f->first, alpha1=f1->first;

     for(;;){
      if(f->first > f1->first){
                    if(f1!=f1_end){
                                    A[alpha1] = i-i1;
                    if(++f1!=f1_end) {alpha1=f1->first;
                    i1=f1->second;}
                    else{ f++; A.erase(alpha);
                        if(f==f_end)break;
                        else{i=f->second; alpha=f->first;}
                    }
            }else {
                        A[alpha] = i-i1;
                        if(f!=f_end && ++f!=f_end)
                        { i=f->second; alpha=f->first; }
            elsebreak;
                        }
      }else if(f->first < f1->first){
                        if(f!=f_end){
                                    A[alpha] = i-i1;
                                    if(++f!=f_end)
{ i=f->second; alpha=f->first;}
                        else{ f1++; A.erase(alpha1);
                            if(f1==f1_end)break;
                            else {alpha1=f1->first;
                            i1=f1->second;}
                        }
                        }else {
                            A[alpha1] = i-i1;
                            if(f1!=f1_end && ++f1!
=f1_end){alpha1=f1->first; i1=f1->second;}
                    else break;
                            }
      } else {
            A[alpha] = i - i1;
            if(f1!=f1_end && ++f1!=f1_end){alpha1=f1->
            first; i1=f1->second;}
            if(f!=f_end && ++f!=f_end){ alpha=f->first;
            i=f->second; }
            if(f1==f1_end && f==f_end)break;
            }
    }
    return * this;
}
```

```
/*
 Definition of operation "*=" between fuzzy numbers.
This operator evaluates the representation of two fuzzy
numbers and updates a result that represents the sum of
the fuzzy numbers.
 */
template<class T> inline FuzzyNumber<T>& FuzzyNumber<T>::
 operator*=(constFuzzyNumber<T>& a)
    {
    typedef map<T,Interval<T> >::const_iterator CI;

    CI f1_end = a.A.end(), f1 = a.A.begin(),
       f_end = A.end(), f = A.begin();

     Interval<T> i=f->second, i1=f1->second;
     T alpha=f->first, alpha1=f1->first;

      for(;;){
        if(f->first > f1->first){
                    if(f1!=f1_end){
                                      A[alpha1] = i*i1;
                       if(++f1!=f1_end) {alpha1=f1->first;
                       i1=f1->second;}
                       else{ f++; A.erase(alpha);
                            if(f==f_end)break;
                            else{i=f->second; alpha=f->first;}
                       }
                }else {
                         A[alpha] = i*i1;
                         if(f!=f_end && ++f!=f_end){ i=f->
                         second; alpha=f->first; }
                  elsebreak;
                       }
          }elseif(f->first < f1->first){
                         if(f!=f_end){
                                        A[alpha] = i*i1;
                                        if(++f!=f_end)
{ i=f->second; alpha=f->first;}
                         else{ f1++; A.erase(alpha1);
                            if(f1==f1_end)break;
                            else {alpha1=f1->first;
                            i1=f1->second;}
                       }
                         }else {
                             A[alpha1] = i*i1;
                             if(f1!=f1_end &&
```

```
++f1!=f1_end){alpha1=f1->first; i1=f1->second;}
                    elsebreak;
                            }
        }else {
                A[alpha] = i * i1;
                if(f1!=f1_end && ++f1!=f1_end){alpha1=f1->
                first; i1=f1->second;}
                if(f!=f_end && ++f!=f_end){ alpha=f->first;
                i=f->second; }
                if(f1==f1_end && f==f_end)break;
                }
        }
    return * this;
}


/*
 Definition of operation "/=" between fuzzy numbers.
This operator evaluates the representation of two fuzzy
numbers and updates a result that represents the sum of
the fuzzy numbers.
*/
template<class T> inline FuzzyNumber<T>& FuzzyNumber<T>
:: operator/=(const Fuzzy Number<T>& a)
   {
   typedef map<T,Interval<T> >::const_iterator CI;

   CI f1_end = a.A.end(), f1 = a.A.begin(),
      f_end = A.end(), f = A.begin();

    Interval<T> i=f->second, i1=f1->second;
    T alpha=f->first, alpha1=f1->first;

      for(;;){
       if(f->first > f1->first){
                    if(f1!=f1_end){
                                    A[alpha1] = i/i1;
                    if(++f1!=f1_end) {alpha1=f1->first;
                    i1=f1->second;}
                    else{ f++; A.erase(alpha);
                        if(f==f_end)break;
                        else{i=f->second; alpha=f->first;}
                    }
            }else {
                    A[alpha] = i/i1;
                    if(f!=f_end && ++f!=f_end){ i=f->
                    second; alpha=f->first; }
```

```
                    elsebreak;
                          }
        }elseif(f->first < f1->first){
                         if(f!=f_end){
                                        A[alpha] = i/i1;
                                        if(++f!=f_end)
{ i=f->second; alpha=f->first;}
                      else{ f1++; A.erase(alpha1);
                         if(f1==f1_end)break;
                         else {alpha1=f1->first;
                         i1=f1->second;}
                 }
                         }else {
                               A[alpha1] = i/i1;
                               if(f1!=f1_end &&
++f1!=f1_end){alpha1=f1->first; i1=f1->second;}
                 elsebreak;
                          }
        }else {
              A[alpha] = i / i1;
              if(f1!=f1_end && ++f1!=f1_end){alpha1=f1->
              first; i1=f1->second;}
              if(f!=f_end && ++f!=f_end){ alpha=f->first;
              i=f->second; }
              if(f1==f1_end && f==f_end)break;
              }
        }
    return * this;
}



/*
Definition of binary operator "+" between fuzzy numbers.
This operator evaluates the representation of two fuzzy
numbers  and returns a result that represents the sum
of the fuzzy numbers.
*/
template<class T> FuzzyNumber<T> operator+(const
FuzzyNumber<T>& a,  constFuzzyNumber<T>& b)
    {
            FuzzyNumber<T> r = a;
             return r+=b;
    }
```

```
/*
```
Definition of binary operator "-" between fuzzy numbers.
This operator evaluates the representation of two fuzzy
numbers and returns a result that represents the sum
of the fuzzy numbers.
```
*/
```
template<class T> FuzzyNumber<T> **operator-(const**
FuzzyNumber<T>& a, const FuzzyNumber<T>& b)
```
        {
                FuzzyNumber<T> r = a;
                return r-=b;
        }
```

```
/*
```
Definition of binary operator "" between fuzzy numbers.*
This operator evaluates the representation of two fuzzy
numbers and returns a result that represents the sum
of the fuzzy numbers.
```
*/
```
template<class T> FuzzyNumber<T> **operator*(const**
FuzzyNumber<T>& a, const FuzzyNumber<T>& b)
```
        {
                FuzzyNumber<T> r = a;
                return r*=b;
        }
```

```
/*
```
Definition of binary operator "/" between fuzzy numbers.
This operator evaluates the representation of two fuzzy
numbers and returns a result that represents the sum
of the fuzzy numbers.
```
*/
```
template<class T> FuzzyNumber<T> **operator/(const**
FuzzyNumber<T>& a,   **const** FuzzyNumber<T>& b)
```
        {
                FuzzyNumber<T> r = a;
                return r/=b;
        }
```

```
/*
```
Functional representation. Given an alpha-level x, this
operator returns a pair (alpha-level, interval) whose
alpha-level is the superior extreme of the alpha-level
set less than x.
```
*/
```

```
template<class T> pair<T,Interval<T> > FuzzyNumber<T>::
operator()(const
  T& x) const
    {
      map<T,Interval<T> >::const_iterator
      p=A.lowerbound(x);
      if(p->first==x){
                        pair<T,Interval<T> >
                        f(p->first,p->second);
                          return f;
      }else{
            p--;
            pair<T,Interval<T> > f(p->first,p->second);
             return f;
      }
    }
```

```
/*
Definition of the operator "<<". It formats output for
the out channel.
*/
```

```
template<class T> ostream& operator << (ostream& os,
const FuzzyNumber<T>& x)
    {
      typedef map<T,Interval<T> >::const_iterator CI;

      os << "{ ";
       for (CI p = x.A.begin(); p!=x.A.end(); p++)
          os << "(" << p->first << ", " << p->second
          << ") ";
       return os << "}";
      }
```

```
/*
Definition of the operator ">>". It gets formatted input
from a channel.
*/
template<class T> istream& operator >> (istream& s,
FuzzyNumber<T>& x) {
 FuzzyNumber<T> z;
 Interval<T> Int;
  char c=0;
 T lev;

 z.A.clear();
```

```
 s >> c;
 if (c == '{')
     while ((s>>c) && c!=`}'){
           if (c==`('){
               s >> lev >> c;
               if (c==`,'){
                   s >> Int >> c;
                   if (c!=`)' || !z.Insert(lev,Int))
                   {s.clear(ios::badbit);break;}
                   else continue;
               }else{ s.clear(ios::badbit);break;}
           }else{ s.clear(ios::badbit);break;}
     }
 else s.clear(ios::badbit);

 if (s && c==`}') x=z;
 else s.clear(ios::badbit);

 return s;
}


/*
Constructor to transform a scalar in a fuzzy number.
It links algebraic arithmetic and fuzzy arithemetic.
*/

template<class T> FuzzyNumber<T>::FuzzyNumber(T& x){
   T zero=T();
   A[zero]=x;
   }


/*
Scalar assignment to a fuzzy number. This operator
links algebraic arithmetic and fuzzy number arithmetic.
*/
template<class T> FuzzyNumber<T>& FuzzyNumber<T>::
 operator=(const T& a)
  {
    typedef map<T,Interval<T> >::const_iterator CI;

    CI f=A.begin(),f_end = A.end();
    T zero=T();

    A.erase(f,f_end);
    A[zero]=a;
```

```
    return * this;
    }


/*
Definition of the operator "+=" between a fuzzy number
and a scalar.

*/
template<class T> FuzzyNumber<T>& FuzzyNumber<T>::
 operator+=(const T& a)
    {
      typedef map<T,Interval<T> >:: iterator CI;

      CI f,f_end = A.end();

        for(f=A.begin(); f!=f_end; f++) f->second+=a;
        return * this;
    }



/*
Definition of the  operator "-=" between a fuzzy number
and a scalar.
*/
template<class T> FuzzyNumber<T>& FuzzyNumber<T>::
 operator-=(const T& a)
    {
      typedef map<T,Interval<T> >::const_iterator CI;

      CI f,f_end = A.end();

        for(f=A.begin(); f!=f_end; f++) f->second-=a;
        return * this;
    }

/*
Definition of the operator "*=" between a fuzzy number
and a scalar. NOTA: OPERAZIONE NON CONSERVATIVA.
*/
template<class T> FuzzyNumber<T>& FuzzyNumber<T>::
 operator*=(const T& a)
    {
      typedef map<T,Interval<T> >::iterator CI;

      CI f,f_end = A.end();
```

```
     for(f=A.begin(); f!=f_end; f++) f->second*=a;
     return * this;
  }

/*
Definition of operator "/=" between a fuzzy number
and a scalar.
*/
template<class T> FuzzyNumber<T>& FuzzyNumber<T>::
 operator/=(const T& a)
  {
    typedef map<T,Interval<T> >::const_iterator CI;

    CI f,f_end = A.end();

     for(f=A.begin(); f!=f_end; f++) f->second/=a;
     return * this;
  }

/*
Definition of the operator "+" between a fuzzy number
and a scalar.
*/
template<class T> FuzzyNumber<T> operator+(const
FuzzyNumber<T>& a, constT& b)
    {
              FuzzyNumber<T> r = a;
               return r+=b;
    }

/*
Definition of the operator "-" between a fuzzy number
and a scalar.
*/
template<class T> FuzzyNumber<T> operator-(const
FuzzyNumber<T>& a, const T& b)
    {
              FuzzyNumber<T> r = a;
               return r-=b;
    }

/*
Definition of the operator "-" between a fuzzy number
and a scalar.
*/
```

```cpp
template<class T> FuzzyNumber<T> operator*(const
FuzzyNumber<T>& a, constT& b)
    {
            FuzzyNumber<T> r = a;
             return r*=b;
    }

/*
Definition of the operator "/" between a fuzzy number
and a scalar.
*/

template<class T> FuzzyNumber<T> operator/(const
FuzzyNumber<T>& a, constT& b)
    {
            FuzzyNumber<T> r = a;
             return r/=b;
    }

/*
Definition of the operator "+" between a scalar and a
fuzzy number.
*/

template<class T> inline FuzzyNumber<T> operator+(const
T& b, const FuzzyNumber<T>& a)
    {
         return a + b;
    }

/*
Definition of the operator "-" between a scalar and a
fuzzy number.
*/

template<class T> inline FuzzyNumber<T> operator-(const
T& b, const FuzzyNumber<T>& a)
    {
      typedef map<T,Interval<T> >::const_iterator CI;

      CI f,f_end = a.A.end();

      FuzzyNumber<T> r;

       for(f=a.A.begin(); f!=f_end; f++) r.A[f->first]
       = b - f->second;
```

```
        return r;

    }

/*
Definition of the operator "*" between a scalar and a
fuzzy number.
*/
template<class T> inline FuzzyNumber<T> operator*(const
T& b, const FuzzyNumber<T>& a)
    {
        return a * b;
    }

/*
Definition of the operator "/" between a scalar and a
fuzzy number.
*/
template<class T> inline FuzzyNumber<T> operator/(const
T& b,constFuzzyNumber<T>& a)
  {
    typedef map<T,Interval<T> >::const_iterator CI;

    CI f,f_end = a.A.end();

    FuzzyNumber<T> r;

    for(f=a.A.begin(); f!=f_end; f++) r.A[f->first]
    =b/f->second;

    return r;
}

/*
A weighting function for summation of intervals with
respect to their alpha-levels.
*/
  double g( double alpha){ return alpha+.5; }

/*
  Definition of fuzzy distance.
*/
template<class T> T FuzzyDist(const FuzzyNumber<T>& a,
const FuzzyNumber<T>& b){
    typedef map<T,Interval<T> >::const_iterator CI;
```

```
CI f1_end = b.A.end(),
   f_end = a.A.end(),
   f1 = b.A.begin(),
   f = a.A.begin();

stack<pair<T,T> > d; // < level,distance>
pair<T,T> dtmp;
T fdist=T();
 double tmpg,tmph;

 while(f!=f_end && f1!=f1_end)
 {
  if(f->first > f1->first)
              {
      d.push(pair<T,T>(f1->first,dist(f->second,
      f1->second)));
               if(f1!=f1_end) f1++;
              }
   else if(f->first < f1->first)
                  {
      d.push(pair<T,T>(f->first,dist(f->second,
      f1->second)));
                     if(f!=f_ond) f++;
                  }
   else {
         d.push(pair<T,T>(f->first,dist(f->second,
         f1->second)));
         if(f1!=f1_end) f1++;
         if(f!=f_end) f++;
         }
 }

 dtmp=d.top();
 fdist += pow(dtmp.second,2.)*
          (g(1.) + (tmpg=g(dtmp.first))) *
          (1.-(tmph=dtmp.first))/2.;
 d.pop();

  while (d.size()){
       dtmp=d.top();
       fdist+= pow(dtmp.second,2.)*
       (tmpg+g(dtmp.first))*(tmph-dtmp.first)/2.;
       tmpg=g(dtmp.first); tmph=dtmp.first;
       d.pop();
  }
```

```
        return sqrt(fdist);
}

/*
    Overtaking between fuzzy numbers. "f1 > f2"
*/
template<class T> T FuzzyOver(const FuzzyNumber<T>& a,
const FuzzyNumber<T>& b){
    typedef map<T,Interval<T> >::const_iterator CI;

    CI f=a.A.begin(),f_end = a.A.end();
    stack<pair<T,T> > o; // <level,override>
    pair<T,T> stmp;
    T   s,h1,h2;

      while (f!=f_end){
            s = (f->second.upper()-b(f->first).
                second.lower()) /
                (f->second.upper()-f->second.lower());
            if(s>1.) s=1.;
            if (s>.0) { o.push(pair<T,T>(f->first,s)); f++;}
            else{break;}
      }

    if (f==f_end) o.push(pair<T,T>(1.,.0));
    else if (s<=.0) o.push(pair<T,T>(f->first,.0));

    s=T(); stmp=o.top(); h1=stmp.first;
      while (o.size()){
                    stmp=o.top();
                    h2=stmp.first;
                    s+=stmp.second*(h1-h2)*(g(h1)
                        +g(h2))*.5;
                    h1=h2; o.pop();
      }
      return s;
}

/*
    Definition of mean interval of a fuzzy number.
*/
template<class T> Interval<T> meanint(FuzzyNumber<T> a){

  typedef map<T,Interval<T> >::const_iterator CI;
  CI f,f_end = a.A.end();
  T l=T(), r=T(), prevl=T(), prevu=T(), preva=T();
```

```
  for(f=a.A.begin()++;  f!=f_end;  f++){
                  l+=((f->second.lower()+prevl)*
                  (f->first-preva));
                  r+=((f->second.upper()+prevu)*
                  (f->first-preva));
                  prevl=f->second.lower();
                  prevu=f->second.upper();
                  preva=f->first;
  }

  l+=(prevu-prevl)*(1-preva);
  r+=(prevu-prevl)*(1-preva);

  Interval<T>  i(l/2.,r/2.);

    return i;
  }

  /*
     Definition of centroid for a fuzzy number.
  */
  template<class T> double centroid(FuzzyNumber<T> a){

  typedef map<T,Interval<T> >::const_iterator CI;
  CI f,f_end = a.A.end();

  T l=T(), r=T(), prevl=T(), prevu=T(), preva=T(), l=T(),
  u=T();

    for(f=a.A.begin()++;  f!=f_end;  f++){
                      l+=(f->second.lower()*f->first
                      +prevl*preva)*(f->second.lower()
                      - prevl);
                l+=(f->first+preva)*(f->second.lower()
                - prevl);

                r+=(f->second.up()*f->first+prevu*preva)
                *(prevu - f->second.up());
                r+=(f->first+preva)*(prevu - f->second.up());


                      prevl=f->second.lower();
                      prevu=f->second.upper();
                      preva=f->first;
    }
```

```
l+=((prevl+prevu)/2.+prevl*preva)*(prevu-prevl)/2.;
l+=(1-preva)*(prevu-prevl)/2.;

r+=((prevl+prevu)/2.+prevl*preva)*(prevl-prevu)/2.;
r+=(1-preva)*(prevl-prevu)/2.;

  return (l+r)/(l+r);
}
```

9.2   FUZZY QUERY SYSTEM. C++ CODE

```
#include<iostream>
#include<stdio.h>
#include<fstream>
#include<string>
#include"BSplineFuzzy.h"

 int main( int argc, char *argv[])
{

 //   Definition of variables, input...

  try{
    BSplineFuzzy< double> B(grado,xg,yg,vliv,x,y,f,EPS);

    B.write(st);


  double l[2],u[2], xgs, ygs,lvl;
      l[0]=(*(*B).u)[1];
      l[1]=(*(*B).v)[1];
      u[0]=(*(*B).u)[(*(*B).u).size()-(*B).M];
      u[1]=(*(*B).v)[(*(*B).v).size()-(*B).M];

// Writing down a map at level lvl
lvl = ...

 for(xx=l[0]; xx<u[0]; xx+=xstep)
  for(yy=l[1]; yy<u[1]; yy+=ystep)
  tomap << xx << " " << yy << " " << (*B)(xx,yy)(lvl).
  second.lower() << " " << (*B)
(xx,yy)(lvl).second.upper()<< " n";

 FuzzyNumber<double> sogliaF(soglia);
```

```cpp
  double levf=...
  double levi=...
  double levs=...

 for(xx=l[0]; xx<u[0]; xx+=xgs){
  for(yy=l[1]; yy<u[1]; yy+=ygs){

      for(double s=lvlf; s>=lvli;s-=lvls){
        FuzzyNumber<double> soglia = FuzzyNumber
        < double>(s);
        // cout << xx << " " << yy << " " << FuzzyOver
        ((*B)(xx,yy), sogliaF) << " " <<
over << endl;
        if(FuzzyOver((*B)(xx,yy),soglia)>over)break;
     }
    tomap << s << " ";
 }
 tomap<< "\n";
}


 } catch(BSplineFuzzy< double>::NoSolution& er){
  cout <<"Numerical Run-Time error number: "
  << er.icaso<< "\n";
  system("PAUSE");
   return -1;
 }

   return 0;
}
```

References

1. D. Altman, Fuzzy set theory approaches for handling imprecision in spatial analysis, *International Journal of Geographical Information Systems* **8** (1994), 271–289.
2. A. Anile, S. Deodato, and G. Privitera, Implementing fuzzy arithmetic, *Fuzzy Sets and Systems* **72** (1995), no. 2, 239–250.
3. A. Anile, B. Falcidieno, G. Gallo, M. Spagnuolo, and S. Spinello, Modeling uncertain data with fuzzy b-splines, *Fuzzy Sets and Systems* **113** (2000), no. 3, 397–410.
4. A. M. Anile and S. Spinella, Fuzzy modeling of sparse data, *International Symposium on Spatial Data Handling* (P. Fisher, ed.), Springer, 2004, pp. 163–172.
5. M. A. Anile, P. Furno, G. Gallo, and A. Massolo, A fuzzy approach to visibility maps creation over digital terrains, *Fuzzy Sets and Systems* **135** (2003), no. 1, 63–80.
6. T. Bailey and A. Gatrell, *Interactive spatial data analysis*, Longman Scientific & Technical, Essex, 1995.
7. H. Bandemer and A. Gebhardt, Bayesian fuzzy kriging, *Fuzzy Sets and Systems* **112** (2000), 405–418.
8. A. Bardossy, I. Bogardi, and W. Kelly, Kriging with imprecise (fuzzy) variograms - II: Application, *Mathematical Geology* **22** (1990), 81–94.
9. R. E. Barnhill, K. Opitz, and H. Pottmann, Fat surfaces: A trivariate approach to triangle based interpolation on surfaces, *Computer Aided Geometric Design* **9** (1992), no. 5, 365–378.
10. C. Bertoluzza, N. Corral, and A. Salas, On a new class of distances between fuzzy numbers, *Mathware & Soft Computing* **2** (1995), 71–84.
11. R. P. Botchen, D. Weiskopf, and T. Ertl, Texture-based visualization of uncertainty in flow fields, *Proceedings IEEE Vis2005*, 2005, pp. 647–654.
12. R. Brown, Animated visual vibrations as an uncertainty visualization technique, *Proceedings of 2nd International Conference on Computer Graphics and Interactive Techniques in Australasia and South East Asia*, 2004, pp. 84–89.
13. J. C. Burkill, Functions of intervals, *Proceedings of the London Mathematical Society* **22** (1924), 375–446.
14. P. Burrough, *Geographic objects with indeterminate boundaries*, pp. 3–28, Taylor & Francis, London, 1996.
15. P. Burrough, P. Van Gaans, and R. MacMillan, High-resolution landform classification using fuzzy k-means, *Fuzzy Sets and Systems* **113** (2000), 37–52.
16. P. Burrough and R. MacDonnell, *Principles of geographical information systems*, Oxford University Press, Oxford, 1998.
17. A. Cedilnik and P. Rheingans, Procedural annotation of uncertain information, *Proceedings IEEE Vis2000*, 2000, pp. 77–84.
18. Y. Chou, *Exploring spatial analysis in geographic information systems*, OnWord Press, New York, 1997.
19. A. Clemantis, M. Spagnuolo, and G. Privitera, *Calcolo di funzioni fuzzy su architetture parallele*, Technical Report, IMA-CNR, Genova, 1995.
20. A. Cohen and N. Gotts, *The 'egg-yolk' representation of regions with indeterminate boundaries*, Taylor & Francis, London, 1996.
21. G. F. Corliss, *Tutorial on validated scientific computing using interval analysis*, see http://www.eng.mu.edu/corlissg/PARA04/READ_ME.html, June 20–23 2004.

22. C. R. Johnson and A. R. Sanderson, A next step: Visualizing errors and uncertainty, *IEEE Computer Graphics and Applications* **23** (2003), no. 5, 6–10.
23. S. M. Debanne, R. A. Bielefeld, G. M. Cauthen, T. M. Daniel, and D. Y. Rowland, Multivariate Markovian modeling of tubercolosis: Forecast for the United States, *Emerging Infectious Diseases* **6** (2000), 148–157.
24. C. DeBoor, On calculating with b-splines, *Journal of Approximation Theory* **6** (1972), 50–62.
25. P. Diamond, Fuzzy kriging, *Fuzzy Sets and Systems* **33** (1989), 315–332.
26. W. M. Dong and F. S. Wong, Fuzzy weighted averages and implementation of the extension principle, *Fuzzy Sets and Systems* **21** (1987), 183–199.
27. D. Dubois and H. Prade (eds.), *Fundamentals of fuzzy sets*, Plenum Press, New York, 1988.
28. D. Dubois and H. Prade, Fuzzy sets and statistical data, *European Journal of Operational Research* **25** (1984), 345–356.
29. D. Dubois, and H. Prade, *Fundamentals of Fuzzy Sets*. Kluwer Academic Press, Inc., Boston, 2001.
30. D. Dubois, E. Kerre, R. Mesiar, and H. Prade, *Fuzzy interval analysis*, Ch. 10, In: *Fundamentals of Fuzzy Sets*, D. Dubbis, H. Prade, (Eds.), Kluwer Academic Press, Boston, 2000.
31. D. Dubois and H. Prade, Fuzzy real algebra: Some results, *Fuzzy Sets and Systems* **2** (1979), no. 4, 327–348.
32. _____, *Possibility theory: An approach to computerized processing of uncertainty*, Plenum Press, New York, 1988.
33. _____, Fuzzy sets, probability and measurement, *European Journal of Operational Research* **40** (1989), 135–154.
34. D. Duff and H. Guesgen, An evaluation of buffering algorithms in fuzzy GISs, *Geographic Information Science, Proceedings of the Second International Conference*, GIScience 2002, Boulder, USA, 2002, pp. 80–92.
35. P. S. Dwayer, *Linear computations*, John Wiley, New York, 1951.
36. M. Ecker, *Geostatistics: Past, present and future*, Encyclopedia of Life Support Systems (EOLSS), Eolss Publishers, Oxford, 2003.
37. G. Edwards, and K. Lowell, Modeling uncertainty in photointerpreted boundaries, *Photogrammetric Engineering & Remote Sensing* **62** (1996), 337–391.
38. M. Egenhofer, E. Clementini, and P. Di Felice Topological relations between regions with holes, *International Journal of Geographical Information Science* **8** (1994), 129–142.
39. M. Erwig and M. Schneider, *Vague regions*, 5th International Symposium on Advances in Spatial Databases, Springer-Verlag, 1997, pp. 298–320.
40. N. M. Patrikalakis et al., Virtual environment for ocean exploration and visualization, *Proceedings of Computer Graphics Technology for Exploration of the Sea*, Rostock, May 1995.
41. P. Fisher, Sorites paradox and vague geographies, *Fuzzy Sets and Systems* **113** (2000), 7–18.
42. R. Fletcher, *Practical methods of optimization (2nd ed.)*, Wiley-Interscience, New York, NY, 1987.
43. C. Fonte, Conversion between the vector and raster data structures using fuzzy geographical entities, *Proceedings of the 7th International Symposium on Spatial Accuracy Assessment in Natural Resources and Environmental Sciences* (M. Painho and M. Caetano, eds.), Instituto Geográfico Português, Lisoba, 2006.
44. C. Fonte and W. A. Lodwick, Areas of fuzzy geographical entities, *International Journal of Geographical Information Science* **18** (2004), no. 2, 127–150.

45. _____, Area, perimeter and shape of fuzzy geographical entities, *Developments in spatial data handling* (P. Fisher ed.), Springer-Verlag, Berlin, 2005, pp. 315–326.

46. _____, Modeling the fuzzy spatial extent of geographical entities, *Fuzzy modeling with spatial information for geographical problems*, Springer-Verlag, Berlin, 2005, pp. 121–142.

47. H. Franssen, A. Eijnsbergen, and A. Stein, Use of spatial prediction techniques and fuzzy classification for mapping soil pollutants, *Geoderma* **77** (1997), 243–262.

48. G. Gallo, I. Perfilieva, M. Spagnuolo, and S. Spinello, Geographical data analysis via mountain function, *International Journal of Intelligent Systems* **14** (1999), 359–373.

49. N. Gershon, Visualization of an imperfect world, *IEEE Computer Graphics and Applications* **18** (1998), no. 4, 43–45.

50. G. Gonçalves and J. Santos, *Propagation of DEM uncertainty: An interval arithmetic approach*, XXII International Cartographic Conference, Spain, 2005.

51. G. Grigoryan and P. Rheingans, Probabilistic surfaces: Point based primitives to show surface uncertainty, *Proceedings IEEE Vis2002* (Boston, MA), 2002, pp. 147–153.

52. H. Guesgen, J. Hertzberg, R. Lobb, and A. Mantler, Buffering fuzzy maps in GIS, *Spatial Cognition & Computation* **3** (2003), 207–222.

53. E. R. Hansen, *Publications related to early interval work of R. E. Moore*, http://interval.louisiana.edu/Moores_early_papers/bibliography.html (2001).

54. E. Hansen, *Global optimization using interval analysis*, Dekker, New York, 1992.

55. E. Hisdal, Are grades of membership probabilities? *Fuzzy Sets and Systems* **25** (1988), 325–348.

56. F. Höppner, F. Klawonn, R. Kruse, and T. Runkler, *Fuzzy cluster analysis*, John Wiley & Sons, Chichester, 1999.

57. M. Jackson and P. Woodsford, *Geographical information systems, GIS Data Capture* Hardware and Software, Longman Scientific & Technical, London, 1991.

58. K. D. Jamison and W. A. Lodwick, The construction of consistent possibility and necessity measures, *Fuzzy Sets and Systems* **132** (2002), no. 1, 1–10.

59. M. Katinsky, *Fuzzy set modeling in geographic information systems*, Master's thesis, Department of Geography, University of Wisconsin-Madison, 1994.

60. A. Kauffman and M. M. Gupta, *Introduction to fuzzy arithmetic: Theory and applications*, Van Nostrand Rheinold, New York, 1991.

61. F. Klawonn, Fuzzy sets and vague environment, *Fuzzy Sets and Systems* **66** (1994), no. 2, 207–221.

62. G. Klir and B. Yuan *Fuzzy sets and fuzzy logic — theory and applications*, Prentice Hall PTR, Englewood Cliffs, NJ, 1995.

63. P. Lancaster and K. Salkauskas, *Curve and surface fitting. An introduction*, Academic Press, London, 1986.

64. S. K. Lodha, A. Pang, R. E. Sheehan, and C. M. Wittenbrink, Uflow: Visualizing uncertainty in fluid flow, *VIS '96: Proceedings of the 7th conference on Visualization '96 (Los Alamitos, CA, USA)*, IEEE Computer Society Press, 1996, pp. 249–ff.

65. S. K. Lodha, C. M. Wilson, and R. E. Sheehan, Listen: Sounding uncertainty visualization, *VIS '96: Proceedings of the 7th Conference on Visualization '96 (Los Alamitos, CA, USA)*, IEEE Computer Society Press, 1996, pp. 189–ff.

66. W. A. Lodwick, *Accuracy of spatial databases, Developing confidence limits on errors of suitability analyses in geographic information systems modeling*, pp. 69–78, Taylor & Francis, London, United Kingdom, 1989.

67. _____, *Constraint interval arithmetic*, Technical Report No. 138, UCD/CCM, 1999.

68. _____, *Interval and fuzzy analysis: A unified approach*, in the series titled Advances in Imagining and Electronic Physics (P. W. Hawkes, ed.), Volume 147, Elsevier Press, Amsterdam, Neatherlands, 2007.

69. W. A. Lodwick and J. Santos, Constructing consistent fuzzy surfaces from fuzzy data, *Fuzzy Sets and Systems* **135** (2003), no. 2, 259–277.

70. W. A. Lodwick and Jorge Santos, Constructing consistent fuzzy surface from fuzzy data, *Fuzzy Sets and Systems* **135** (2003), no. 2, 259–277.

71. W. A. Lodwick, W. Munson, and L. Svoboda, Attribute error and sensitivity analysis of map operations in geographic information systems: Suitability analysis, *International Journal of Geographic Information Systems* **4** (1990), no. 4, 413–428.

72. L. Tran and L. Duckstein, Multiobjective fuzzy regression with central tendency and possibilistic properties, *Fuzzy Sets and Systems* **130** (2002), 21–31.

73. D. Luenberger, *Linear and nonlinear programming*, Addison-Wesley Publishing, Reading, MA, 1984.

74. R. MacMillan, W. Pettapiece, S. Nolan, and T. Goddard, A generic procedure for automatically segmenting landforms into landform elements using DEMs, heuristic rules and fuzzy logic, *Fuzzy Sets and Systems* **113** (2000), 81–109.

75. A. McBratney and I. Odeh, Application of fuzzy sets in soil science: Fuzzy logic, fuzzy measurements and fuzzy decisions, *Geoderma* **77** (1997), 85–113.

76. J. R. Miller, D. C. Clibum, J. J. Feddema, and T. A. Slocum, Modeling and visualizing uncertainty in a global water balance model, SAC '03: *Proceedings of the 2003 ACM Symposium on Applied Computing* (New York, NY, USA), ACM Press, 2003, pp. 972–978.

77. R. Moore, *Interval analysis*, Prentice Hall, Englewood Cliffs, NJ, 1966.

78. R. E. Moore, *Automatic error analysis in digital computation*, Technical Report LMSD-48421, Lockheed Missile and Space Division, Sunnyvale, CA, 1959.

79. _____, *Interval arithmetic and automatic error analysis in digital computing*, Technical Report Ph.D. thesis, Stanford University, 1962.

80. _____, *The dawning*, *Reliable Computing* **5** (1999), no. 4, 423–424.

81. R. E. Moore, W. Strother, and C. T. Yang, *Interval integrals*, Technical Report Space Div. Report LMSD703073, Lockheed Missile and Space Division, Sunnyvale, CA, 1960.

82. R. E. Moore and C. T. Yang, *Interval analysis I*, Technical Report Space Div. Report LMSD285875, Lockheed Missile and Space Division, Sunnyvale, CA, 1959.

83. Archimedes of Siracusa, *On the measurement of the circle*, Cambridge University Press, Cambridge, 1897.

84. C. Olston and J. D. Mackinlay, Visualizing data with bounded uncertainty, INFOVIS '02: *Proceedings of the IEEE Symposium on Information Visualization (InfoVis'02) (Washington, DC, USA)*, IEEE Computer Society, 2002, p. 37.

85. A. T. Pang, C. M. Wittenbrink, and S. K. Lodha, Approaches to uncertainty visualization, *The Visual Computer* **13** (1997), 370–390.

86. G. Privitera, M. A. Anile, and S. Deodato, Implementing fuzzy arithmetic, *Fuzzy Sets and Systems* **72** (1995), no. 2, 239–250.

87. A. Rosenfeld, The diameter of a fuzzy set, *Fuzzy Sets and Systems* **13** (1984), 241–246.

88. Santos, Lodwick J., W., and Neumaier, A new approach to incorporate uncertainty in terrain modelling, Lecture Notes in Computer Science, *Proceedings of the Second International Conference on Geographic Information Science: GIScience 2002 (Boulder, Colorado)*, Springer-Verlag, February 2002, pp. 291–299.

89. W. Silvert, Fuzzy indices of environmental conditions, *Ecological Modelling* **130** (2000), 111–119.

90. J. Stolfi and L. de Figueiredo, *Self-validated numerical methods and applications*, Institute for Pure and Applied Mathematics (IMPA), Rio de Janeiro, 1997, Monograph for the 21st Brazilian Mathematics Colloquium (CBM'97), IMPA.
91. T. Sunaga, Theory of an interval algebra and its application to numerical analysis, *RAAG Memoirs* **2** (1958), 547–564.
92. I. Turksen and T. Bilgiç, Measurement of membership functions: Theoretical and empirical work, *Fundamentals of fuzzy sets*, The Handbook of Fuzzy Sets Series, Kluwer Academic Publishers, Boston, 2000, pp. 195–230.
93. C. Tomlin, *Geographic information systems and cartographic modeling*, Prentice Hall, New York, 1990.
94. L. Tran and L. Duckstein, Comparison of fuzzy numbers using a fuzzy distance measure, *Fuzzy Sets and Systems* **130** (2002), no. 3, 331–341.
95. L. Usery, *A conceptual framework and fuzzy set implementation for geographic features*, pp. 71–85, Taylor & Francis, London, 1996.
96. Z. Wang and G. J. Klir, *Fuzzy measure theory*, Plenum Press, New York, 1992.
97. M. Warmus, Calculus of approximations, *Bulliten de l'Académié Polonaise de Sciences Cl. III* **4** (1956), 253–259.
98. C. M. Wittenbrink, IFS fractal interpolation for 2D and 3D visualization, VIS '95: *Proceedings of the 6th Conference on Visualization '95 (Washington, DC, USA)*, IEEE Computer Society, 1995, p. 77.
99. J. D. Wood and P. F. Fisher, Assessing interpolation accuracy in elevation models, *IEEE Computer Graphics and Applications* **13** (1993), no. 2, 48–56.
100. K. L. Wood, K. Otto, and E. K. Antonsson, Engineering design calculations with fuzzy parameters, *Fuzzy Sets and Systems* **52** (1992), 1–20.
101. H. Q. Yang, H. Yao, and J. D. Jones, Calculating functions of fuzzy numbers, *Fuzzy Sets and Systems* **55** (1993), no. 3, 273–283.
102. R. C. Young, The algebra of many-valued quantities, *Mathematische Annalen Band* **104** (1931), 260–290.
103. L. A. Zadeh, Fuzzy sets, *Information and Control* **8** (1965), 338–353.
104. ———, Probability measures of fuzzy events, *Journal of Mathematical Analysis and Applications* **23** (1968), 421–427.
105. ———, The concept of a linguistic variable and its application to approximate reasoning: Part I, *Information Sciences* **8** (1975), 199–249.
106. ———, The concept of a linguistic variable and its application to approximate reasoning: Part II, *Information Sciences* **8** (1975), 301–357.
107. ———, The concept of a linguistic variable and its application to approximate reasoning: Part III, *Information Sciences* **9** (1975), 43–80.
108. ———, Fuzzy sets as a basis for a theory of possibility, *Fuzzy Sets and Systems* **1** (1978), 3–28.
109. J. Zhang and R. Kirby Alternative criteria for defining fuzzy boundaries based on fuzzy classification of aerial photographs and satellite images, *Photogrammetric Engineering & Remote Sensing* **65** (1999), 1379–1387.
110. J. Zhang and N. Stuart, Fuzzy methods for categorical mapping with image-based land cover data, *International Journal of Geographical Information Science* **15** (2001), 175–195.
111. H. J. Zimmermann, *Fuzzy set theory and its applications*, Kluwer-Nijhoff Publishing, Boston, 1986.

Index

Milton Keynes UK
Ingram Content Group UK Ltd.
UKHW040054071024
449327UK00019B/545